Iron Melting Cupola Furnaces
For the Small Foundry

Steve Chastain

**Iron Melting Cupola Furnaces
For the Small Foundry**

By Stephen D. Chastain

First Edition
3rd Printing
with additional material

ISBN 0-9702203-0-8

3 4 5 6 7 8 9 0

Table of Contents

WARNING – DISCLAMER

This book is to provide information on the methods used by the author to develop a cupola. The author is neither an engineer nor scientist. No liability is assumed for the use of the information contained in this book. If you do not wish to be bound by the above, you may return the book for a full refund.

Warning: As with any fire, the incomplete combustion of coke produces carbon monoxide (CO), a poisonous gas. Carbon monoxide burns off harmlessly to carbon dioxide in air. The flame at the top of the cupola is the burning carbon monoxide. The auto-ignition temperature of CO is about 1200 degrees F. If you use a higher stack for better charge preheat, the discharge temperature may be as low as 400 degrees. In such case it may be desirable to install a gas pilot to be sure the gas ignites upon leaving the cupola, and or you may maintain a boxer fan in the work area to provide additional ventilation. A regenerative blast preheater was deliberately left out of the book because of the hazards of pumping CO about the work area.

MATH: The math used in this book is no more complicated than basic high school algebra. Any one should be able to do the computations with a $9 calculator and a half-hour of practice (read the instructions for the calculator). As you read the book you should work the examples on the calculator. You will see just how easy it is. You will be missing out on the ability to customize your cupola if you don't. Get over your math fears and do it. Good luck!

FOREWORD

Since I wrote *"BUILDING SMALL CUPOLA FURNACES"* five years ago, I have heard back from many people building them in both this country and other places around the world. Probably the most successful of any of these was Steve Chastain. He approached this project in a very scientific manner and his first results more than justified his great efforts; he had perfectly hot usable iron on his very first heat! I was so impressed with Steve's results that I encouraged him to put his work on paper for the benefit of the rest of us. He has now done so, and I know you will be very pleased with his thorough book and drawings. If you are building a cupola of about 10 in bore, you will find meticulous instructions that allow you to duplicate Steve's own amazing furnace; and I am sure you will agree this book is a terrific bargain. Happy and safe melting to all.

Stewart Marshall
17th May, 2000

About the Author:

I grew up in central Florida, and have always been considered a creative person. While attending Florida Southern College, I studied economics, music, and theater. Later, I started my own band that toured nationally for several years. While on tour, I noticed the inadequacy of the commercially available production equipment. I designed and built my own line of staging equipment, later forming SDC Electronics to manufacture and sell the equipment. I left the music and staging businesses to pursue a career in engineering. At the time of this writing, I am entering my junior year as a mechanical engineering student. While in school, I have been working as a lighting consultant, rebuilding industrial lift trucks, and running a small foundry and machine shop that produces parts for antique engines and equipment.

One of my theatrical dimmers. This rack Contains thirty-six 2.4 kW dimmers

ACKNOWLEDGMENTS

I would like to thank Dave Gingery and Stewart Marshall for convincing me I had something to say. They provided support during the writing process. I want to thank my neighbors, especially Randy Smith and his family for putting up with the perpetual experiment in my yard. Jim Brown, Ph.D. always comes over when I need a hand with the operation of the cupola. Peter Reich introduced me to true cupola authorities, Norm Lilleybeck and Rod Schueler, who reviewed an early manuscript and encouraged me to continue. William Provis of Modern Equipment Company provided additional research material, which also helped confirm my figures. Most of all, I would like to thank my wife, Gina, who endures my long hours and keeps a positive attitude.

PURPOSE:

The purpose of this book is to provide a set of plans for a cupola that will produce 35 – 55 lbs. of iron every 8 to 12 minutes. The cupola will be built from scrap. The cost will be $50 to $200 depending on the type of refractory used. The book will describe construction, setup, and basic operational parameters. Several factors affect operation. Height and size of tuyeres, size of windbelt, location of tap and slag holes, and minimum height of stack are critical for proper operation. Other critical parameters include size and weight of iron and coke charges. The formula must be followed exactly or you will have poor operation. Brief theory and calculations describe larger cupolas, up to 18" in diameter.

This book is not intended to be a foundry manual. The minimum foundry experience required for this project should include competence pouring non-ferrous metals. There are several good books available from Lindsay publications and Stewart Marshall regarding foundry work.

INTRODUCTION:

Old engines fascinate me and I love to restore old machines. Many engines and machines are scrapped because of a broken casting; parts are extremely expensive, or nonexistent. I had several pieces that could be fixed if I could cast the parts. My first project was a set of pistons for an antique four-cylinder generator set. Then came manifolds, waterpumps, magnetos, and cylinder sleeves. As friends found out I could pour iron, several tractor and machine projects followed. Lathe accessories followed by a cam grinder were designed and built. Now there seemed to be no limit to what I could do. I have always felt I could have anything I wanted, I just had to build it. The cupola lets you restore old machines or design new ones. You have a sense of freedom when you can design, cast, and build any piece of equipment you can dream up. Man has been pouring iron for thousands of years. It is not nearly as difficult as you might think. Good luck, I hope you enjoy your cupola as much as I enjoy mine.

Machining my first Piston Casting

This piston went in a 10 kW generator set that powers my house

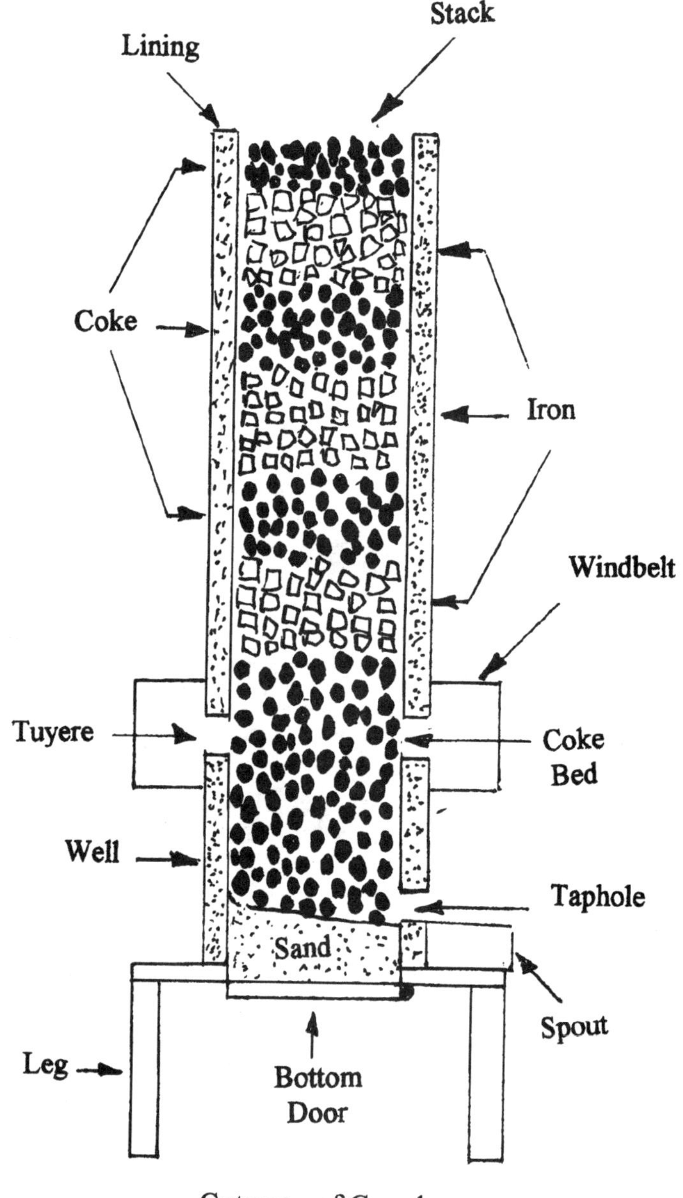

Cutaway of Cupola

I.
CUPOLA THEORY AND DESIGN CONSIDERATIONS:

Cupola Zones:

There are five areas in a cupola. These areas are the well, combustion, reduction, melting, and the preheating areas.

The **well** is at the bottom of the cupola. It stores the liquid iron until the cupola is tapped. Because of the close contact with the coke and exposure time, most of the iron's carbon pick up occurs in the well. The depth of the well affects the temperature of the tapped iron. Hotter iron comes from a shallow well. Slag separates and floats on top of the iron in the well.

The **combustion zone** is where the blast enters the cupola and reacts with the coke to form carbon monoxide and carbon dioxide. The heat of reaction superheats the iron and generates hot gases, which melt and preheat the charge. This is the area where the thermal energy or heat is generated.

The **reduction zone** is outside the combustion or oxidation zone. The iron is superheated in this zone. Oxides of iron are also reduced here.

The **melting zone** is the area from the top of the coke bed to where the iron actually melts. The melting zone is located between the reduction zone and the preheat zone.

The **preheat zone** is above the melting area and extends to the top of the charged material. The temperature of the charge must rise from ambient or room temperature to melting temperature. The charge receives its largest amount of heat gain in this area. Gases entering this region are approximately 2200 degrees Fahrenheit, however after giving up their heat to the charge; they exit at 400 to 800 degrees Fahrenheit. Do not underestimate the importance of the preheat area. Good melting

11

depends on adequate preheat. Other functions of the preheat include the drying of the charge materials.

The cupola stack should be kept full so that charge materials can be adequately preheated. The lower in the stack the charges sit, the lower the melting zone and cooler the metal. The height of the stack is limited by the ability of the coke bed to support the weight of the charges and the ability of the blast and hot gases to penetrate the charges. The efficiency of the cupola is largely dependent on the heat transfer between the rising hot gases and the descending charges. The longer the contact time, the better the heat transfer. Thus a higher bed and stack are superior to a low bed and short stack. Short stacks do not allow sufficient preheat.

Pouring a 60 pound faceplate for a lathe

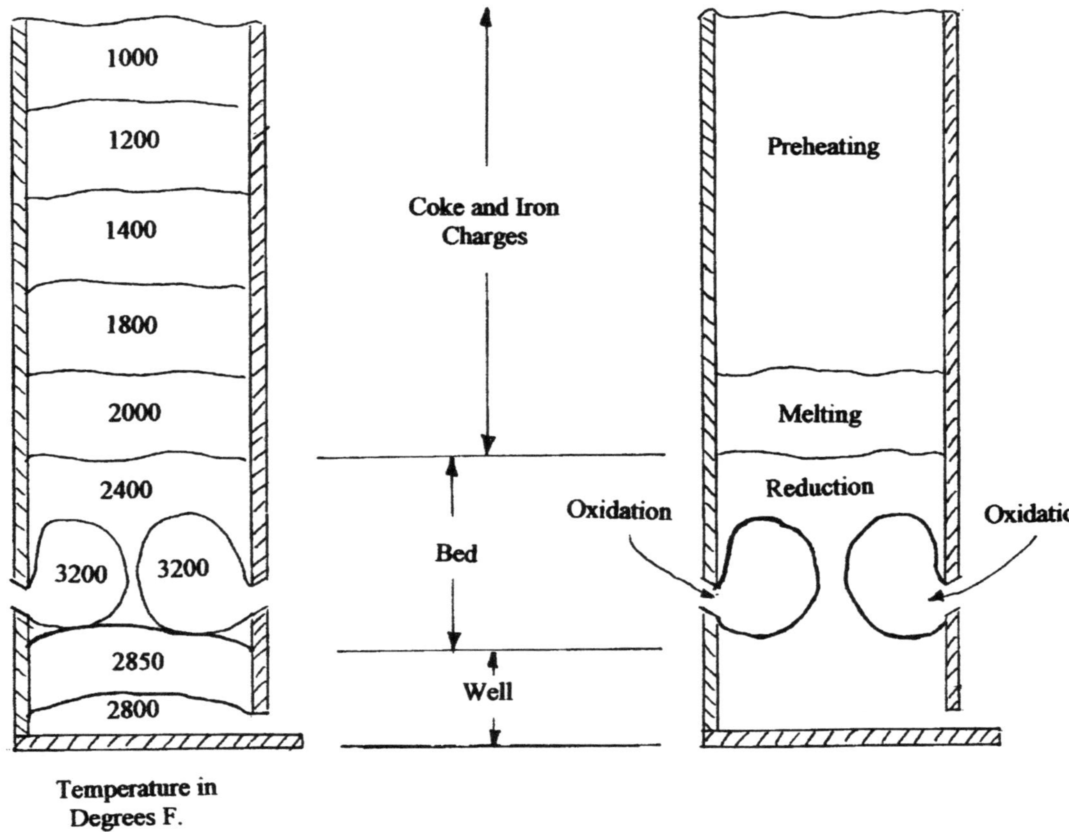

Cupola Zones

13

MATERIAL SIZE:

The size of the metal making up the charge is critical to maintaining efficient preheating and melting. If the pieces are too large, they will not melt in the melt zone but will work their way down to the tuyere level. The iron will not superheat and this could lead to a frozen taphole. If small pieces are charged without some type of grouping, they often burn up or will clog up the voids in the stack. This could lead to bridging in the cupola. Small and thin pieces should be gathered and pressed into a briquette. Borings may be melted by filling short lengths of pipe and plugging the ends A good size for the 10 inch cupola is 2 by 3 by ½ -inch.

Coke must be screened to proper size, being 1/10 to 1/12 the inside diameter of the cupola. The space between larger pieces allows the gasses to rise faster is the stack. This raises the combustion zone thereby reducing the preheat of the charge. Smaller pieces could pack too tightly, increasing resistance to the gas flow. Fines must be eliminated as they end up being blown out of the stack as a shower of sparks (sometimes rising 60 feet or more) or come down as dust covering the whole area. If limestone is used as a desulfurizing flux, it should be in pebble form.   Larger pieces will not decompose properly. However, if it is too fine, it will clog the furnace or be blown out of the stack.

MELT RATIOS:

By varying the iron to coke ratio and the amount of air, any cupola is capable of a wide range of melting rates and melting temperatures. In all cases, the air and coke must be balanced to maintain consistent melting.

When the coke and air are unbalanced, melting problems arise. Too much coke results in a rising bed, wasted coke, slow melting, and lower temperatures. Too much air burns away the bed, lowering the temperature and leaving free oxygen in the cupola to burn the iron. A "sparky tap" indicates too much air.

Ideally, stack gases should contain 13 percent carbon dioxide, 13.2 percent carbon monoxide, and 73.8 percent nitrogen. The balance of coke and air is critical for proper operation. Metal temperature is a direct result of a properly maintained bed height. A proper bed will allow taps at 2750 degrees. However, imbalance in the cupola will lead to lower temperatures.

COMMON CUPOLA RATIOS:

Conventional industrial cupolas are built to melt from 2 to 32 tons per hour. Their ratios offer a place to start the design process. However, the surface area to volume ratio A_s/V is higher for our small cupolas, putting the 10-inch cupola at the higher end of the thermal loss scale. Small cupolas are not as efficient as large ones.

Thermal Balance Sheet:

Of all the energy contained in the coke only 25% to 65% is used to heat the iron charge. The remaining 35% to 75% is lost. The losses are as follows:

	% total heat available in charge		
Lining and Tuyeres	4	to	20
Stack Discharge	6	to	30
CO production (incomplete combustion)	17		17
Slag	8	to	8
TOTAL LOSSES:	35%	to	75%
HEAT LEFT FOR MELTING:	65%	to	25%

Conventional cupolas melt 10 pounds of iron per square inch of area between the tuyeres per hour. The 10-inch cupola could theoretically melt 785 pounds of iron per hour.

Melt rate in $\underline{\text{pounds}}$ $= 10 \; \pi \, (D/2)^2$ pounds / hour
hr

D= diameter

$\pi = 3.14$

D = 10 inches

Melt rate $= 10\pi(10/2)^2 = 785$ pounds / hour

The maximum ratio of iron to coke is 10 to 1; ten pounds of iron per pound of coke. Our 10-inch cupola is melting at a 6 to 1 ratio. Best cupola operation occurs with incomplete combustion. The stack gasses should contain 13% CO_2 (carbon dioxide) 13.2% CO (carbon monoxide) and 73.8 % N (nitrogen). The CO burns to CO_2 as it is discharged from the stack, giving a large visible flame if melting at night.

If our 10-inch cupola is melting 330 pounds per hour, we can calculate the amount of air required. One pound of carbon requires 113 cubic feet of air to provide the 13% CO_2 - 13.2% CO ratio. Coke contains approximately 90% carbon. What are the air requirements for the 10-inch cupola?

$$\frac{330 \text{ pounds iron}}{\text{hour}} \times \frac{1 \text{ pound coke}}{6 \text{ pounds iron}} \times \frac{.9 \text{ pound carbon}}{1 \text{ pound coke}} \times \frac{1 \text{ hour}}{60 \text{ min}} = \frac{.825 \text{ pound C}}{\text{min}}$$

(113 cubic feet air / pound coke) X .825 pound coke = 93.2 cubic feet air

The shop vac used on our cupola is rated at 90 cubic feet per minute. The calculations appear realistic for our situation.

TUYERE AREA

The tuyere area is based on the inside diameter of the cupola at the tuyere level. Standard ratios for smaller cupolas range from ¼ to 1/6 the cross sectional area of the cupola at the tuyeres. Earlier practice favored larger tuyeres that increase in size as they approach the inside of the cupola. The coefficient of discharge is much higher for this type of nozzle. However, modern practice favors smaller tuyeres and higher blast pressures. The theory is, as pressure increases, discharge velocity at the tuyeres increases. Blast penetration of the bed will increase thereby lowering the combustion zone.

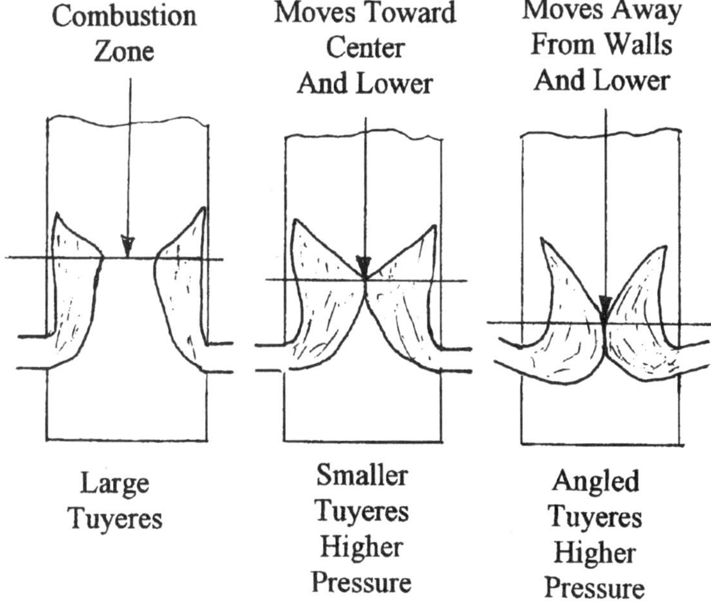

Theory of Blast

As the blast increases the combustion zone will move closer to the center of the cupola, reducing the heat loss through the walls. It should also be possible to further lower the combustion zone by angling the tuyeres downward.

Calculate the tuyere area for the 10-inch cupola using the ratio: 1/6 the sectional area at tuyere level.

$$D = 10 \text{ inches} \qquad A = \pi(D/2)^2$$

$$A = \text{area}$$

$$\pi = 3.14 \qquad A = \pi(10\text{-inches}/2)^2 = 78.5 \text{ inches}^2$$

$$\text{Tuyere area} = 1/6\ A = \frac{1 \times 78.5 \text{ inches}^2}{6} = 13.1 \text{ inches}^2$$

13.1 inches is the required area. Now to find a suitable diameter pipe.

$$\text{area of 2-inch pipe} = \pi(D/2)^2 = \pi(2/2)^2 = 3.14 \text{ inches}^2$$

$$\text{Number of tuyeres} = (13.1 \text{ inches}^2 / 3.14 \text{ inches}^2) = 4.17$$

$$\text{Area of 2 ½-inch pipe} = \pi(2.5 \text{ inches}/2)^2 = 4.9 \text{ inches}^2$$

$$\text{Number of tuyeres} = \frac{13.1 \text{ inches}^2}{4.9 \text{ inches}^2} = 2.66$$

4 tuyeres at 2-inches diameter are closer to the desired ratio than 3 tuyeres at 2 ½-inches diameter. 2-inch pipe will be used.

CUPOLA HEIGHT:

Cupola height (before legs) is determined in two steps. The stack height above tuyeres and the well depth + constant make up the two calculations. For small cupolas a constant of 5-inches will be used. Larger cupolas use 10 inches. The well depth will be considered first. Deep wells increase the difficulty of maintaining hot iron; therefore a conservative (shallow) well will be used for our experimental cupola. The theoretical capacity of the cupola is 1 pound per square inch of sectional area at the tuyeres. For a 10-inch cupola this tap would be approximately 78 pounds. A tap of 35 pounds should certainly be attainable, therefore we will calculate for a 35-pound tap. The density of iron is approximately 4.4 ounces / cubic inch.

$$\text{Ounces of iron} = \frac{35 \text{ pounds iron}}{1} \times \frac{16 \text{ ounces}}{\text{pound}} = 560 \text{ ounces iron}$$

$$\text{Cubic inches of iron} = \frac{560 \text{ ounces of iron}}{4.4 \text{ ounce} / \text{inch}^3} = 127 \text{ inches}^3$$

Well volume available for iron = .46 total well area

$$\text{Total well volume} = \frac{127 \text{ inches}^3}{.46} = 277 \text{ inches}^3$$

$$\text{Height to slag hole (from bottom of well)} = \frac{\text{total well volume}}{\text{area of well}}$$

Area of well = 78.5 inches^2

$$\text{Height to slag hole (from bottom of well)} = \frac{277 \text{ inches}^3}{78.5 \text{ inches}^2} = 3.52 \text{ inches}$$

Height to tuyeres includes the (height of the 2-inch sand bottom)+ (3.52 inch height of the slag hole) + (5-inches (constant))

Height to tuyeres = 2 inches + 3.52 inches + 5 inches = 10.52 inches.

HEIGHT ABOVE TUYERES:

Height above tuyeres is a simple calculation. As shown earlier higher stacks are superior to shorter stacks. The minimum stack height is 4.5D at tuyere level. Between 6.2 to 6.5 is considered optimum.

Calculate the height above tuyeres using the 4.5D ratio.

$$4.5 \ X \ 10\text{-inches} = 45\text{-inches}$$

The total height of the cupola less legs = sum of both heights + tuyere diameter.

Sum = height to tuyeres + height above tuyeres + tuyeres

Σ = 10.52 inches + 45 inches + 2 inches = 57.52 inches

Height of cupola without legs = 57.5 inches

Minimum leg height = length of door + 6 inches + depth of sand bed covering the foundation. Leg height may be adjusted upward from this figure to increase operator comfort.

The ratio (1.3 X total tuyere area) gives minimum sectional area of the windbelt, (1.75 X tuyere area) would be the maximum for this type of cupola.

CALCULATION OF THE BED HEIGHT:

Because any cupola can be operated over a wide melting rate, actual bed height depends upon the operation of the cupola. For larger cupolas the height is given by:

$$6 + 10.5 \sqrt{\frac{\text{blast pressure in inches}}{1.73}}$$

The constant of 6-inches is added to the minimum height and represents the maximum height of the bed. The constant may be adjusted; however a higher bed usually will give hotter iron. The minimum height of the bed for the 10" cupola using 5" blast pressure is calculated as follows:

$$10.5 \sqrt{\frac{5 \text{ inches}}{1.73}} = 17.85 \text{ inches}$$

CALCULATION OF CHARGE WEIGHTS:

The weight of coke used per charge is estimated by:

$.0787 \text{ lbs./inch}^2 \times \text{ Sectional Area inch}^2$

The weight of iron charges is proportional to the weight of the coke charges. Common ratios are from 6:1 to 10:1. Calculate the charge weights for the 10" cupola:

Weight of coke: $.0787 \text{ lb./inch}^2 \times 78.5 \text{ inch}^2 = 6.18$ lbs.

Weight of iron at a 6:1 ratio: $6 \times 6.18\text{lbs.} = 37$ lbs. iron

CALUCLATIONS FOR 18 INCH CUPOLA:

$D = 18$ inches

$\text{Area} = 254 \text{ inches}^2$

Maximum theoretical[*] tap $= 254$ pounds

163 pounds = 64% of theoretical tap

163 pounds = 2608 ounces

$$\text{Cubic inches iron} = \frac{2608 \text{ ounces}}{4.4 \text{ ounce/inch}^3} = 592 \text{ inches}^3$$

$$\text{Total well volume} = \frac{592 \text{ inches}^3}{.46} = 1288 \text{ inches}^3$$

$$\text{Height to slaghole} = \frac{\text{total well volume}}{\text{sectional area}}$$

$$\text{Height to slaghole} = \frac{1288 \text{ inches}^3}{254 \text{ inches}^2} = 5.07 \text{ inches}$$

Height to tuyeres = Σ bottom, tuyeres, slaghole, 5 inches

Sand depth on bottom = 3 inches

Height to tuyeres = (3 + 5 + 5.07) inches $\approx$ 13 inches

Tuyere area = .2 sectional area = 51 inches^2

2-½ inch pipe has approximately 4.9 inches^2 area therefore 10 tuyeres would give 19.3 % of sectional area.

Windbelt sectional area = 1.3 to 1.6 tuyere area

1:1.3 = 58 square inches, dimensions 6 by 9 5/8 inches

Height above tuyeres = 4.5 D at tuyeres

$$4.5 \times 18 \text{ inches} = 81 \text{ inches}$$

Total height without legs = 81 inches + 2 ½ inches + well

Height (163 lb. well) = (81 + 2.5 + 13.07) inches = 97 inches

*in modern commercial practice well capacities may be up to 400 lbs.

Well Depths for 18 inch Cupola

18 inch cupola height to slaghole from sand bottom	3 inch sand bottom height to tuyeres from base plate
150 pound well = 4.67 inches	12.67 inches
163 pound well = 5.07 inches	13.07 inches
200 pound well = 6.23 inches	14.23 inches
250 pound well = 7.78 inches	15.78 inches

Use the 6.5 D height ratio and angle the tuyeres downward 15 degrees for the larger wells.

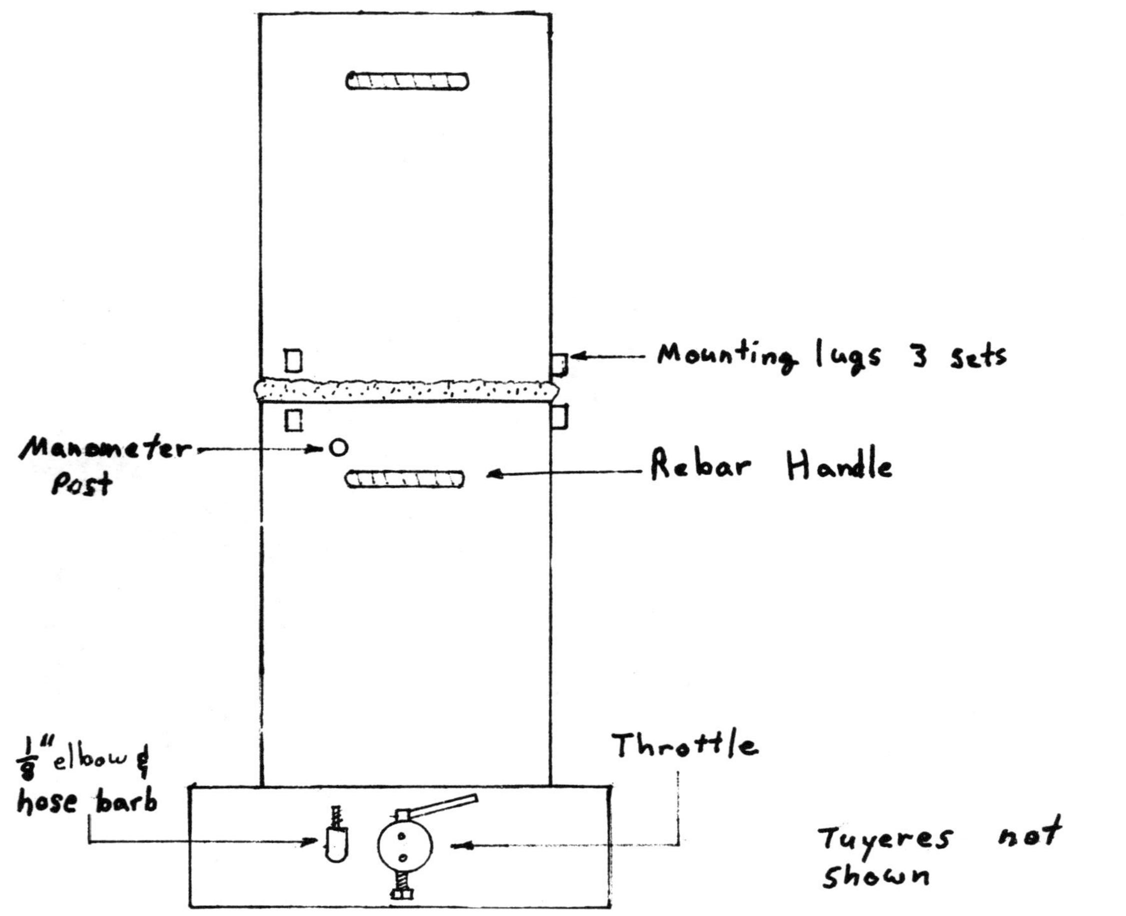

Mounting lugs 3 sets
Rebar Handle
Manometer Post
1/8" elbow & hose barb
Throttle
Tuyeres not shown

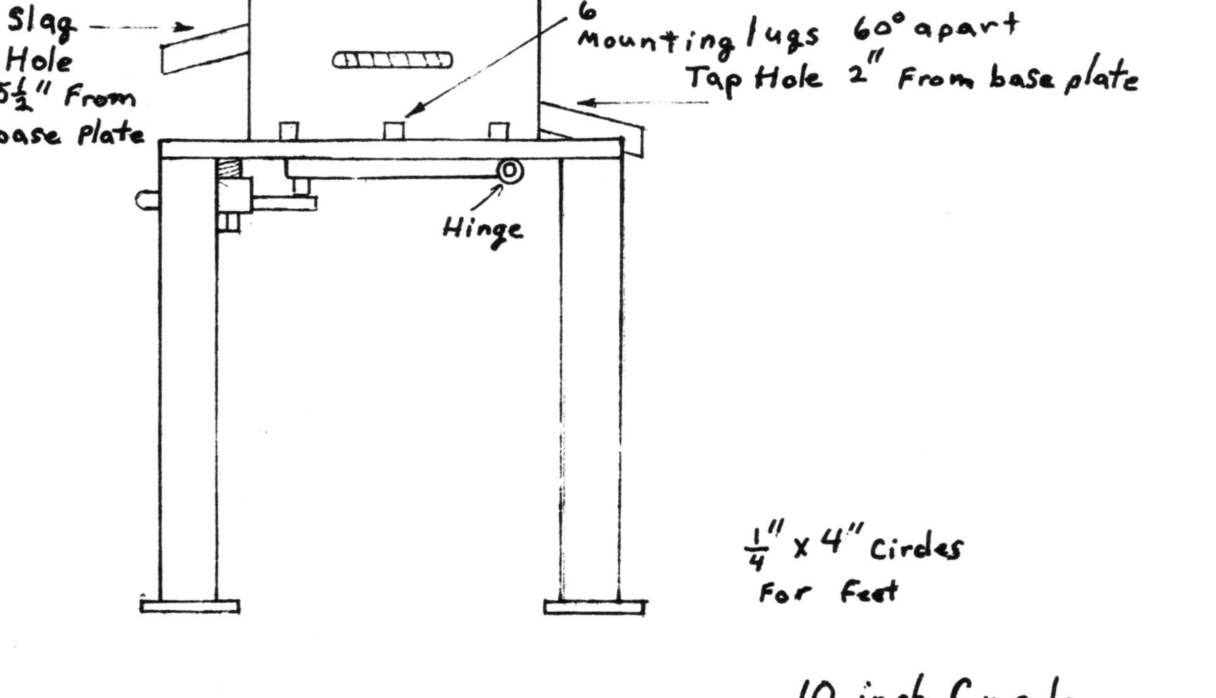

Slag Hole 5½" From base Plate
6 Mounting lugs 60° apart
Tap Hole 2" From base plate
Hinge
¼" x 4" Circles For Feet
10 inch Cupola
Dec. 26 1998

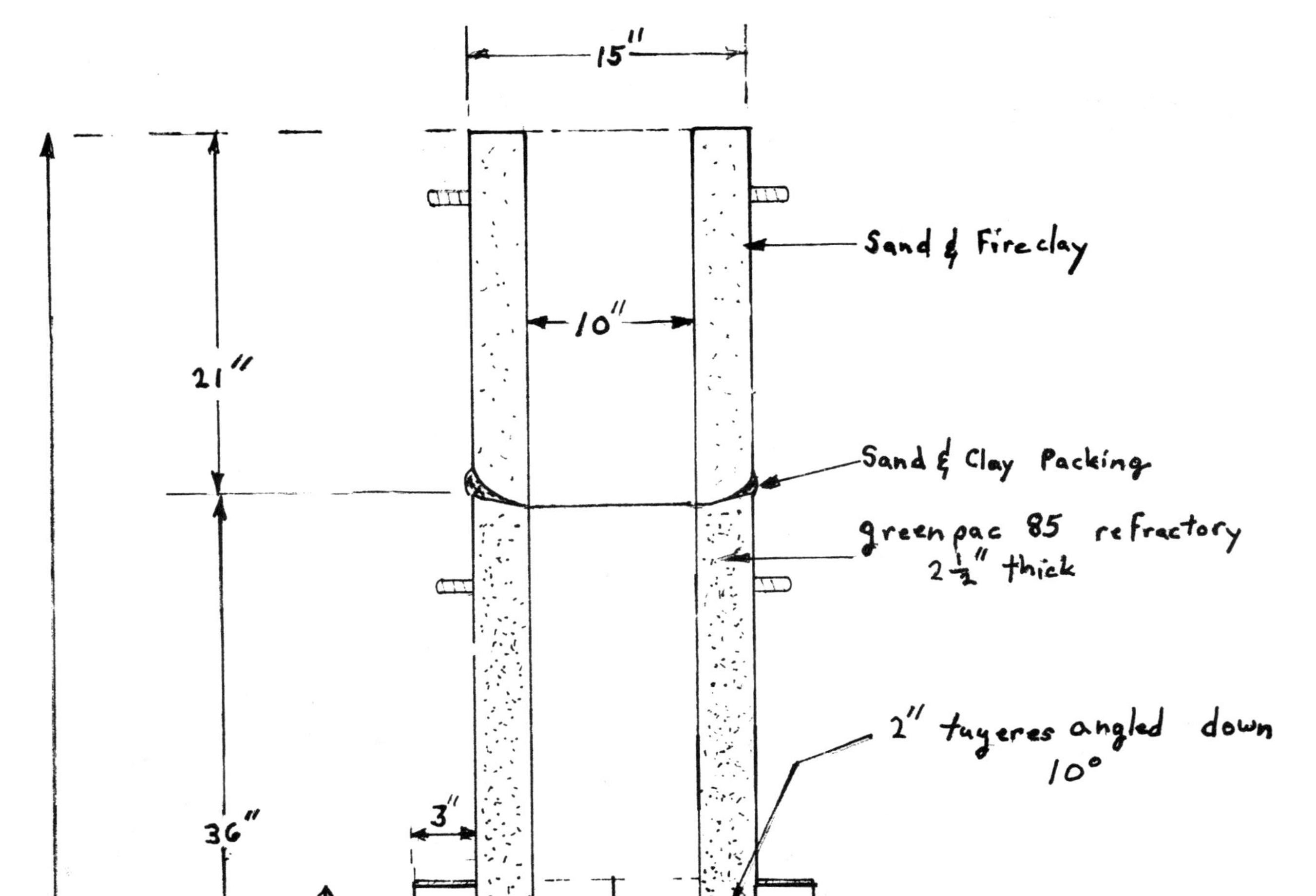

15"
10"
21"
36"
3"
Sand & Fireclay
Sand & Clay Packing
greenpac 85 refractory
2½" thick
2" tuyeres angled down
10°

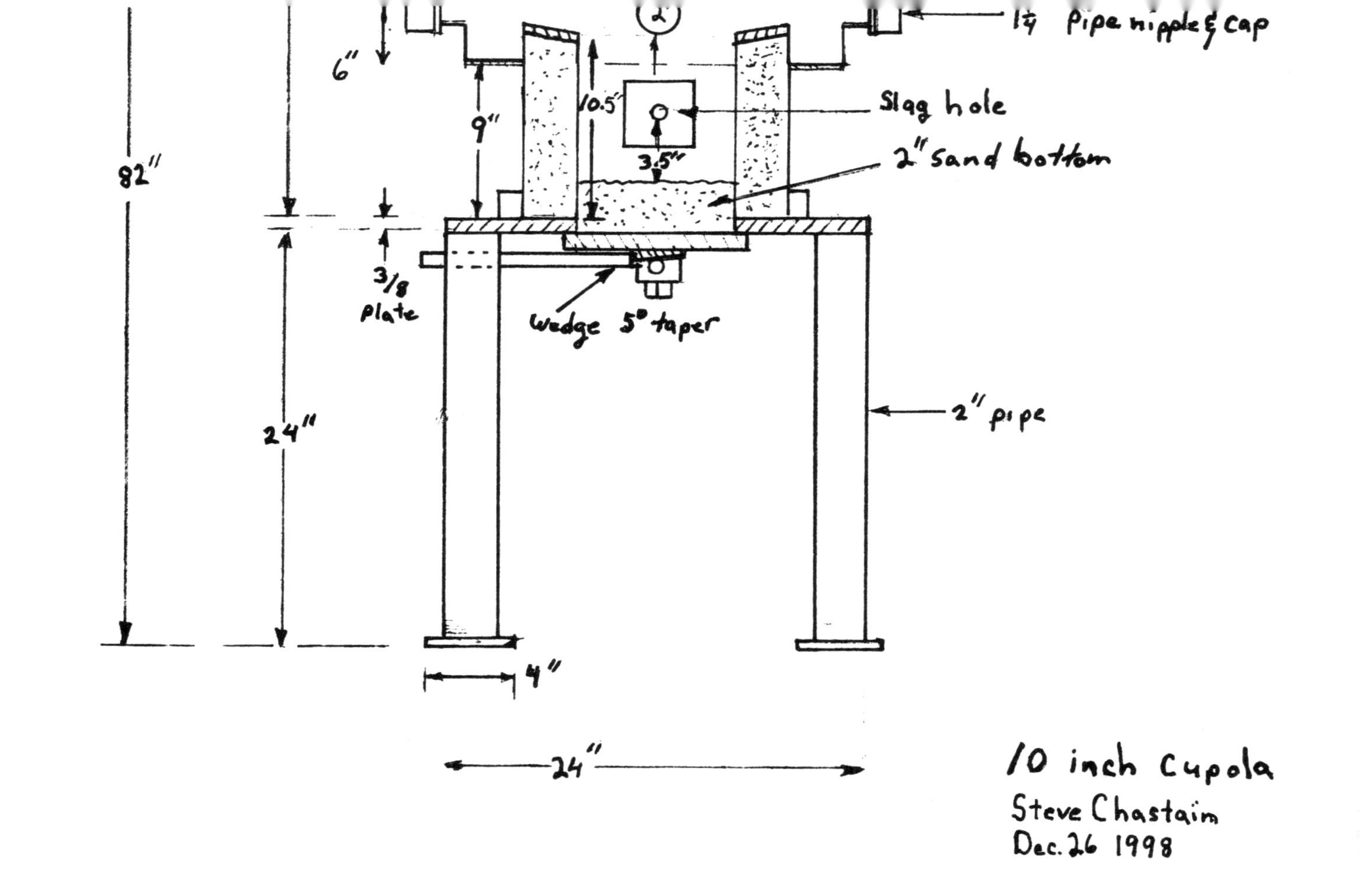

1¼ Pipe nipple & Cap
Slag hole
2" sand bottom
6"
9"
10.5"
3.5"
3/8 plate
Wedge 5° taper
2" pipe
4"
24"
82"
24"
10 inch Cupola
Steve Chastain
Dec. 26 1998

BUILDING A 10 INCH CUPOLA

MAKE A CIRCLE CUTTING ATTACHMENT:

This project requires several precision circles to be cut with a torch. The first thing to do is make a circle cutting attachment. Each torch is different so the drawing is not dimensioned.

The body is made from 1-inch square aluminum stock. I

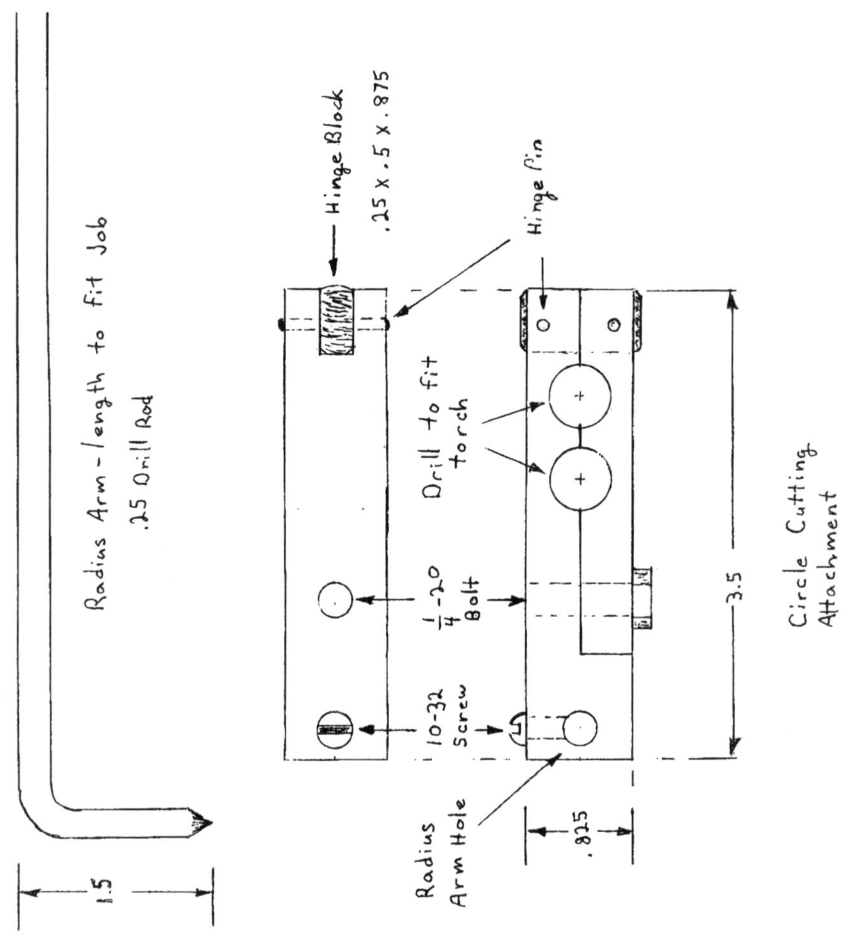

poured my stock while making other aluminum castings. Square the stock and drill the holes for your particular torch, the hinge and screws. Mill or saw out a slot to receive the hinge block. Cut the hinge block from ¼ inch steel plate and drill it to receive the hinge pins. Clamp the body in a vise and cut it open with a hacksaw. Tap the screw holes ¼ - 20 and install the

Completed Cutting Attachment

hinge. Clean up the assembly with a file for a snug fit on your torch. Various radius arms are made from drill rod. Heat the end until red, then bend the right angle. Heat again until non-magnetic, then quench to harden. Clean the tip with emery cloth and dip in oil. Heat the tip until straw brown and quench again, this tempers the tool. When cool, grind the tip to a point. I used ¼" drill rod, however it is not as rigid as I would like; 5/16" rod might be better.

THE CUPOLA SHELL:

The shell should be the first thing you get. It needs to be aired out before you start cutting, so get it first then start looking for other materials.

Cut the shell from discarded 100 lb. propane cylinders. These can be found at any propane supplier. Once the propane cylinders' bottoms become pitted by rust, they are removed

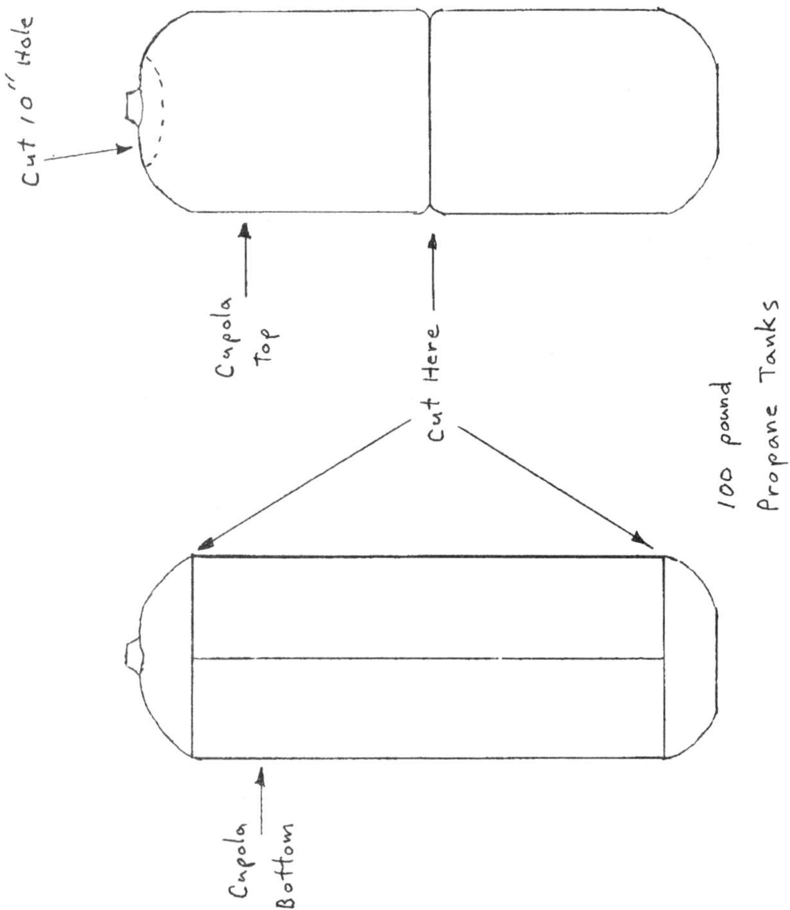

from service. Most gas dealers are glad to get rid of them once they become rusty. While you are there, have them remove the valve. You can also do it at home with a pipe wrench that has a 3-ft. pipe over the handle for leverage.

Cylinders come in 2 types. One is rolled from a single 3-ft. sheet with the top and bottom welded on. (I prefer this type for the bottom) The other type is pressed out of 2 tubes that are then welded together in the middle. I used this type for the top part of the cupola. It took me several months to collect all of the materials from the scrap heap, so my tanks were well ventilated by then.

Any welder knows cutting into a tank can cause an explosion, so always fill the tank with water before cutting it. Cut the top and bottom from one cylinder. Cut the other cylinder around the middle. Save the top half of the second cylinder for the top of the cupola. Plug the valve hole in the short top section and cut a 10-inch hole in the lid using the circle cutting attachment. Clean up the edges with a hand grinder (I use a 4 inch and 9 inch grinder). The 9-inch grinder makes fast work of this job.

To remove the paint from the cylinders, paint them with paint stripper. A single edge razorblade used in a gasket

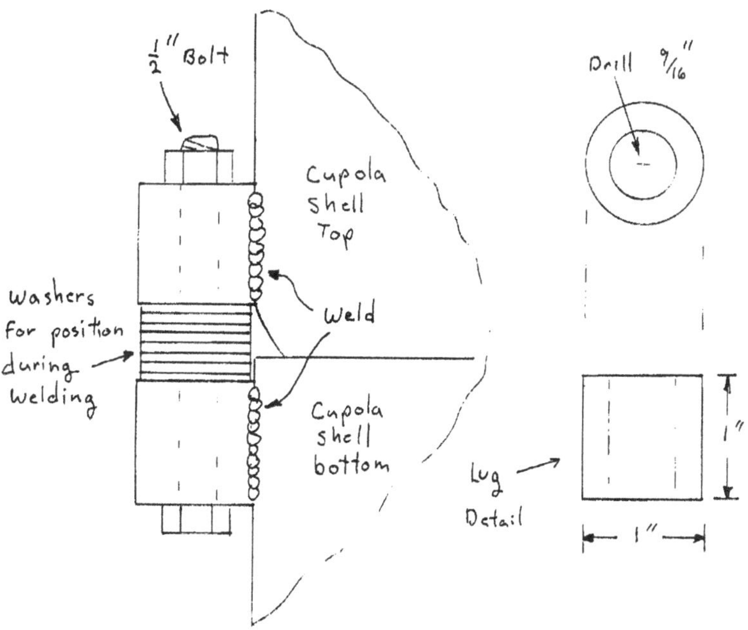

removing tool makes this messy job go much faster. Once these are clean, drill the tap and slag holes 1 inch in diameter. Tuyeres are the openings in the sides of a cupola that admit the blast of air. Locate the tuyere holes so that the bottom of each hole is 11 inches from the base plate. Cut the holes using a hole saw. It will be slightly larger than the 2-inch pipe diameter. That is fine since you will need the room to angle the tuyeres down 10 degrees. Next, build and install the wind belt. When the wind belt is completed, weld the lugs on the bottom of the shell that are used to bolt on the base plate. The six lugs used are made from 1 inch long, 1 inch in diameter rod chucked in the lathe and drilled through the center with a 9/16-inch bit. Weld three more lugs on the top of the shell. Stack the short section on the shell. Bolt another lug separated by washers to each of the 3 lugs you just welded. Weld the new lugs to the top half; these will be used to draw the top half down to form a good seal in the uncured refractory.

Bend rebar handles using a torch to heat the bent-area red hot, quickly clamp in a vise and bend. These were sized to be a comfortable fit to my hand while wearing welding gloves. Weld the handles to the top of the shells. Weld an extra set on the bottom of the lower shell.

THE WIND BELT:

The wind belt wraps around the cupola to deliver air to the tuyeres. Cut the top and bottom of the wind belt from ¼ inch plate using the circle cutting attachment. You could use thinner material as long as you are able to weld it without burning it up. I find myself standing on the wind belt during set up, so I am happier with the thicker stock used here. Once cut, clean them up with a hand grinder. Mark the shell for the top section of the wind belt then turn the shell upside down. Put the top ring of the wind belt over the shell and check the fit. Mark the high spots on the ring, grind them and check the fit again. This could take several Trials. Once fitted, you should be able to pound it into place with a 2-lb. hammer.   Once the first ring is in position, tack weld it into place. The tuyeres should be cut 2 ½ inches long.

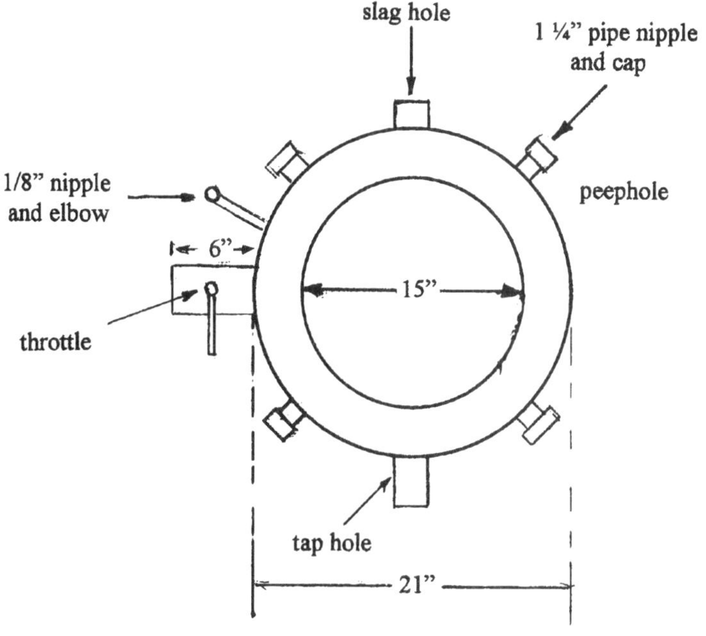

Cupola Top View

Install the tuyeres maintaining the 10-degree angle towards the base plate. The innermost bottoms of the tuyeres should now be 10 ½ inches from the base plate. Be sure to maintain the 10-inch diameter between the tuyeres or you will not be able to fit the lining form in place. You will want to radius the inner edge of each tuyere with a hand grinder before you install them. You can do your final fitting after they are welded in place. Fit the remaining ring and pound it into position maintaining 6 inches between the top and bottom plates. Tack it into place. Measure and mark the locations of the tuyeres on the ring so you can find and locate the peepholes later. The belt is wrapped with strips of

16-gage sheet steel. It is difficult to maintain a perfectly straight strip while welding the wind belt so cut four strips 17 inches by 7 inches. These will be easy to weld and the excess can quickly be removed with a hand grinder. Finally, locate and drill the peepholes with a hole cutting saw. Install the pipe nipples into the peepholes and weld into place. Install the throttle and pressure tap. When you are satisfied and everything is lined up, weld completely around both rings. Clean the job up with a hand grinder.

Make the throttle body from a 6-inch length of 2 ½-inch tubing. Tubing is measured by the outside diameter making it different than pipe. Bore one end to accept the shop vac hose. Alternately, the shop vac hose could be attached with a strip of sheet metal wound around the pipe and hose, the assembly being secured with a pair of hose clamps. Substitute 2-inch pipe with the sheet metal arrangement if tubing can not be found. Later I modified the assembly by welding a flange to the throttle input. I made a flanged adapter for the shop vac that allowed me to attach the blower directly to the cupola, bypassing the hose and eliminating the vacuum's canister. A 15% increase in airflow was recorded by eliminating the hose and canister. I would strongly encourage use of the flanged adapter. You should also be aware that considerable heat is radiated from the cupola, the vacuum will melt if it is too close to the stack. Make a spacer 12 to 18 inches long. Remove the blower from the adapter if you turn it off for more than 1 minute.

The throttle rod is made from ¼ inch by 5-½ inch steel rod. Mill, grind or file a flat on the throttle rod to accept the throttle plate, which is made from thin sheet metal and attached with 2 #6-32 screws. The throttle plate is cut in the shape of a disk that just clears the inside diameter of the throttle body. The action of the throttle is like that of one in a carburator. A spring and a pair of ¼-20 nuts go the bottom of the throttle rod to provide resistance to throttle movement.

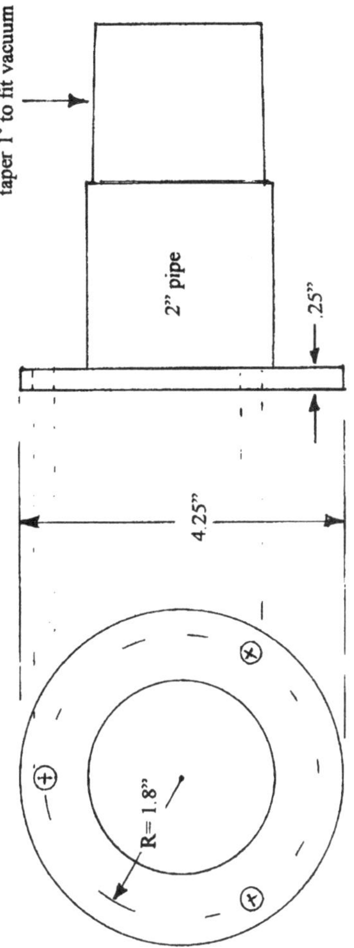

Flanged Adapter

SLAG DOOR:

A 5-inch by 6-inch rectangle is now cut around the slag hole. Four sections of 13/4-inch by 16-gauge strip are bent around the edges of the door and welded into place. Round the corners

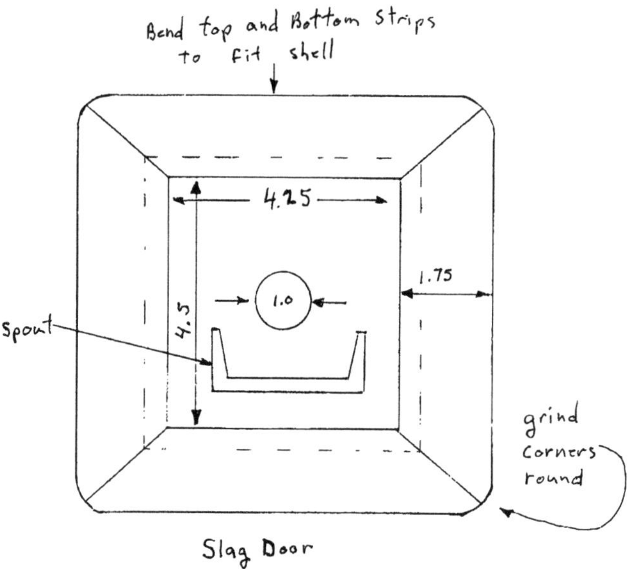

with a grinder. Several 1-inch spikes are made from stiff steel wire and welded on the inside of the door. These spikes help hold the uncured fireclay in place. The slag door is held in place with a 20-inch by 3/8-inch steel rod passing through 2 hooks also made from 3/8-inch steel rod welded to the shell.

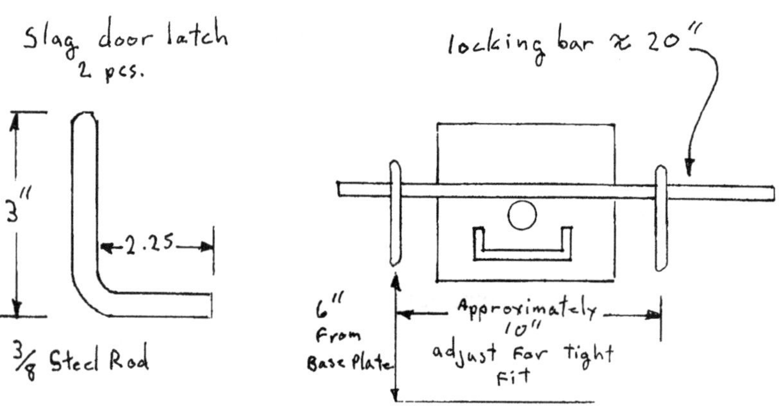

BASE PLATE AND DOOR:

Cut the base plate and door from 3/8-inch plate. This is the minimum thickness. It can be thicker but don't make it thinner. The base plate is cut 24 inches by 24 inches square. This is a change from the original design, so you can catch the drops on a steel sheet. A 10-inch circle is cut in the center of the base plate. The shell is centered over the cut out and the location of the mounting holes are marked and drilled. Cut the door 11 inches in diameter. Drill with ¼ inch holes spaced 1 inch apart, as shown in the photo. These holes are to vent the steam coming from the sand bed. On cold days you can see the steam shooting out of these holes!

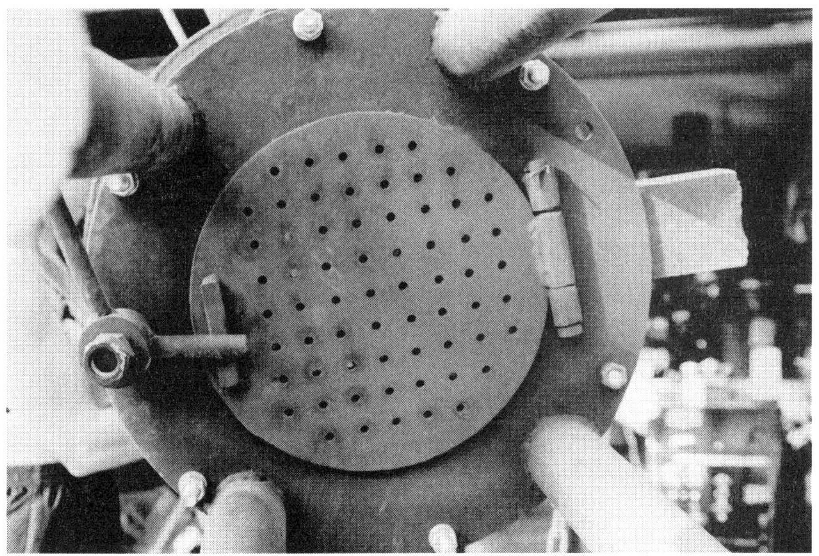

Detail of Bottom Door

Make a hinge pin by turning down a ¾ inch diameter rod until there is .060 clearance between the pin and the inside diameter of heavy wall ½ inch pipe. You could drill a solid rod

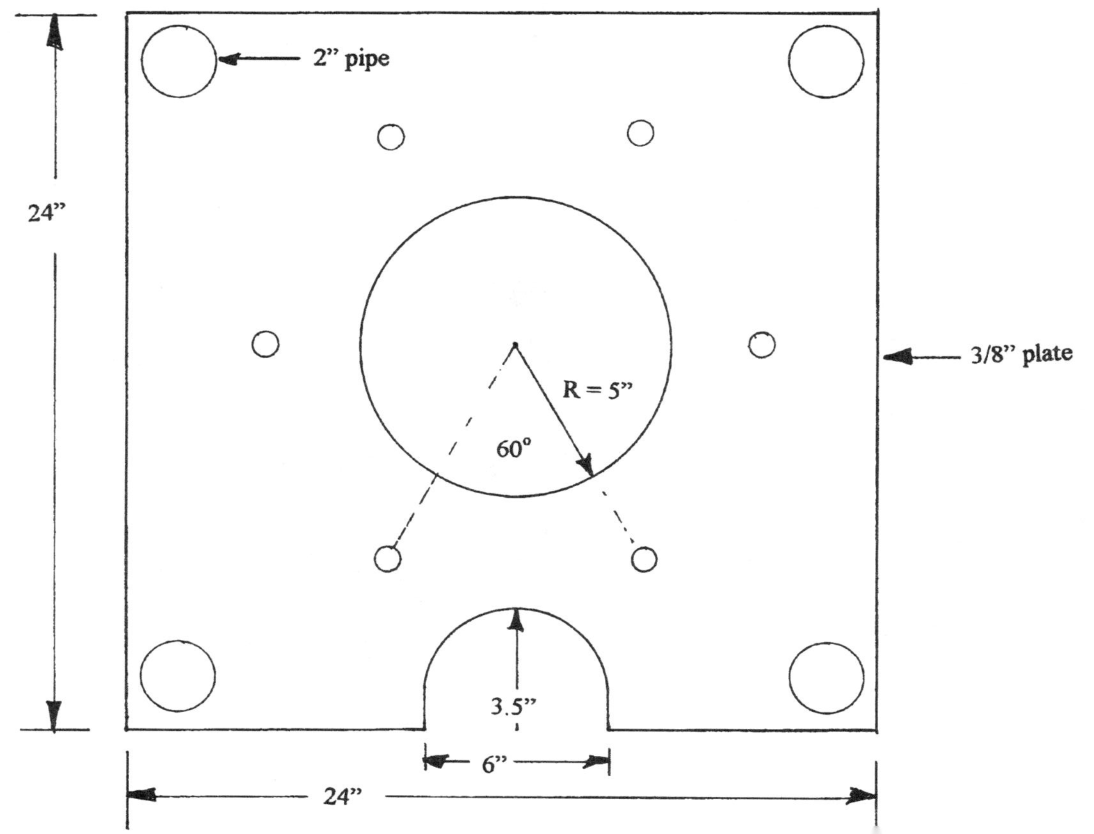

Detail of Modified Baseplate

for the outer portion of the hinge. Be sure to keep the clearance specified so that the hinge will not bind when hot. The hinge must support the weight of all the charges, be sure it is sturdy. Align the door and hinge, then weld it all together.

Cut a wedge with a 5-degree taper from ½ by 3-inch stock. Weld it opposite the hinge. Weld the feet to the legs. Square and weld the legs to the baseplate. At this point you can layout the latch. The angle between the arms will be approximately 100 to 120 degrees. It should latch without hitting the legs. The pivot pin is made from a 4-inch length of ¾ inch screw stock. Make the pivot from a 2-inch piece of 2-inch round stock. Drill the hole for the latch pin and weld the pin in place. Double nuts on the bottom of the latch pin make the assembly adjustable for a perfect fit.

Flip the base plate over and attach the shell with the six ½-inch bolts.

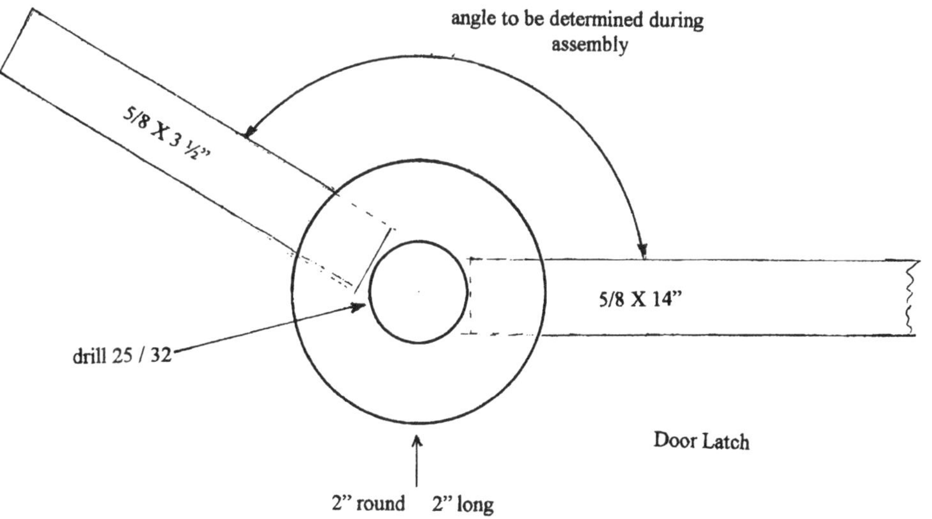

SPOUT:

A spout is made from a 5-inch length of 3-inch channel iron. Hold it in position on the shell and mark the base plate for a relief cut. The cut is made, cleaned up, and the spout is welded in place on the shell. The spout, when lined with sand and fireclay should be below the taphole. This helps keep iron from freezing in the taphole. Ideally you should get a 1-inch drop between the taphole and the spout lining.

LINING THE FURNACE:

The furnace is lined by ramming refractory around a form. The form is located relative to the shell by using four wood strips at 90 degrees to each other. The strips are 2 1/2 inches wide and 36 inches long. Ram the refractory around the form and raise the wood strips as each layer of refractory is laid.

Fireclay and sand are fairly light when compared to commercial refractory. A fireclay mixture can be rammed with a heavy stick or rod, however you will have to use a 1-inch steel rod and a 2-pound hammer to ram some of the commercial refractory. Commercial refractory is expensive but holds up very well. I use a product made by A.P. Green called "Green Pac." It is rated for 3000 degrees. They make "Blue Pac which is rated for 3200 degrees; it is a little more expensive. Others make a similar product. You are looking for 85% alumina phos bonded plastic. This refractory cures extremely hard. I found my refractory under "Firebrick" in the phone book.

Because the top of the furnace is subjected to much lower temperatures it is lined with a sand and fire clay mixture. The mixture is one 50-pound bag of coarse sand, one 50-pound bag of fine sand and one 50-pound bag of fireclay. Pour half a bag of each into a wheelbarrow. Turn it over until mixed then add the rest and turn it over for quite awhile. It must be well mixed, so be patient. When it is well mixed start adding water by misting with a garden hose. Don't let the water puddle in the mixture or the clay will separate from the sand. This process takes a while, so be prepared to do some shoveling. When the mixture has the consistency of stiff mortar it is about right. Don't let the mixture become too wet, it will not cure well and will slough off when fired. When you are satisfied with the consistency, cover it with plastic and take a break for a few hours.

Stewart Marshall Describes is his book *Building Small Cupolas* an alternate way of mixing refractory. He mixes his sand and then adds about half the water. He mixes the sand and water very well and slowly adds the clay while constantly turning. Eventually he works the remaining water into the mix. He claims this method causes the clay to stick to the sand better than the dry mix.

MAKING THE FORM:

Make the form by wrapping 24-gauge sheet metal around two 10-inch plywood discs. These can be cut with a saw, however turning them on a lathe is better. The top disk has a large hole drilled for your fingers; a wood screw is placed in one side of the disc and a 1-inch deep slot is cut in one edge. The bottom disk has a stout screw put in one side. When removing the disk from the cupola you will grab this screw with a pair of pliers and pull it up. Cut a piece of sheet metal 36-inches long. The width of the piece is the circumference of the disc plus 1-inch. For a 10-inch disc, this would be 32 ½ inches wide. Make a right angle bend 1-inch deep and 36-inches long. This will hook into the saw slot of each disk. Tape the sheet metal in place with duct tape. If the metal is springy and difficult to work, roll it up to a diameter less than 10 inches and try taping it again.

After ramming the lining, peel the tape off the top disk, grab the screw with pliers and remove the disk. Remove the bottom disk in a similar manner. Collapse the sheet metal and remove it. Don't be concerned if the form has collapsed a little out of round while ramming the lining. Set the bottom wooden disc back in the top of the furnace and carefully hammer it back down through the bore using the steel rod. This will trim the excess refractory from the high spots. This excess material can

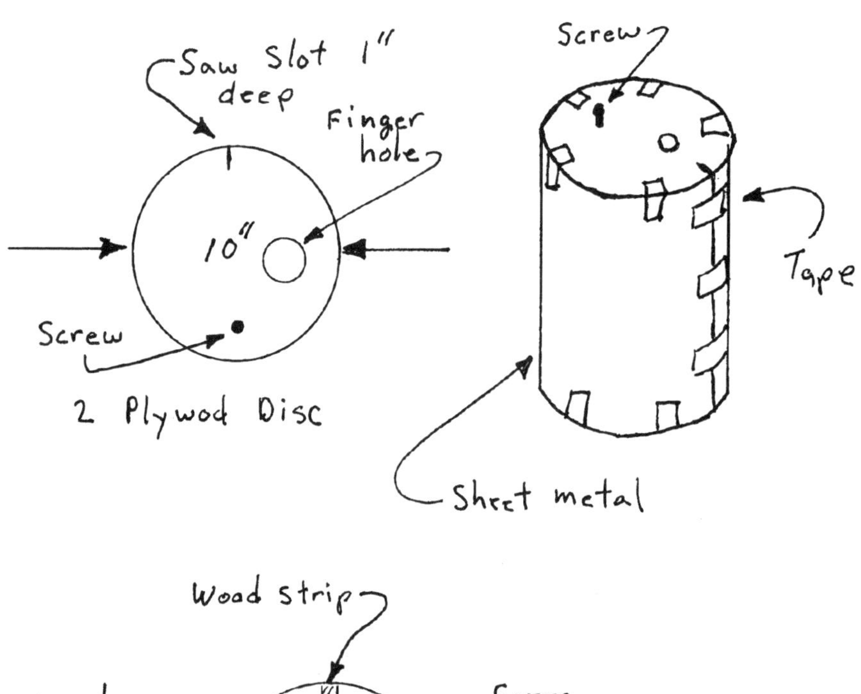

be used to fill any voids. Using a piece of welding rod, poke holes every inch or two around the lining. This is to vent the lining during the curing process. Set the form up in the top half of the cupola shell and repeat the process using the fireclay and sand mixture. Cost savings are the reason for using the fireclay mixture here.

Before ramming the lining, the tap hole void is made by opening the bottom door and inserting a wooden form. Bolt it in place through the tap hole. Ram the lining around the form to

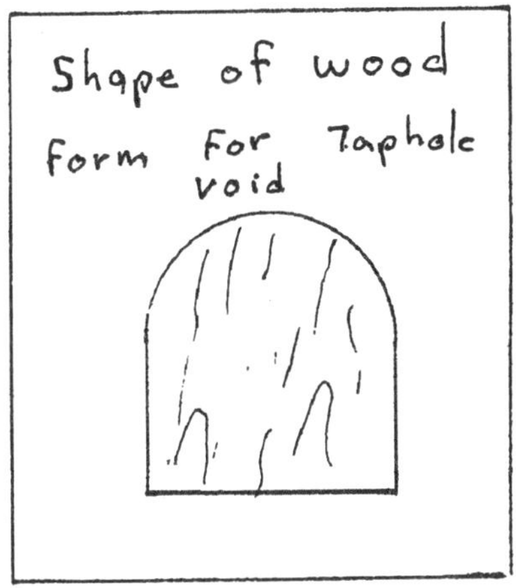

leave a void for the fireclay "taphole." After ramming, the wooden form is removed. Alternately, you could cut the refractory with a knife and a hammer. During operation, this space will be packed with the fireclay mixture. Pushing a piece of ½-inch diameter rebar through from the inside of the cupola will make the taphole. From the outside of the cupola, using a wet finger, the hole is then smoothed to the conical shape as shown. After each heat, the fireclay section is broken out. A new taphole will be made for each heat. This eliminates any iron buildup in the taphole. An iron-coated taphole may cause the taphole to freeze up at the beginning of the melt. It is best to start with a fresh taphole each heat.

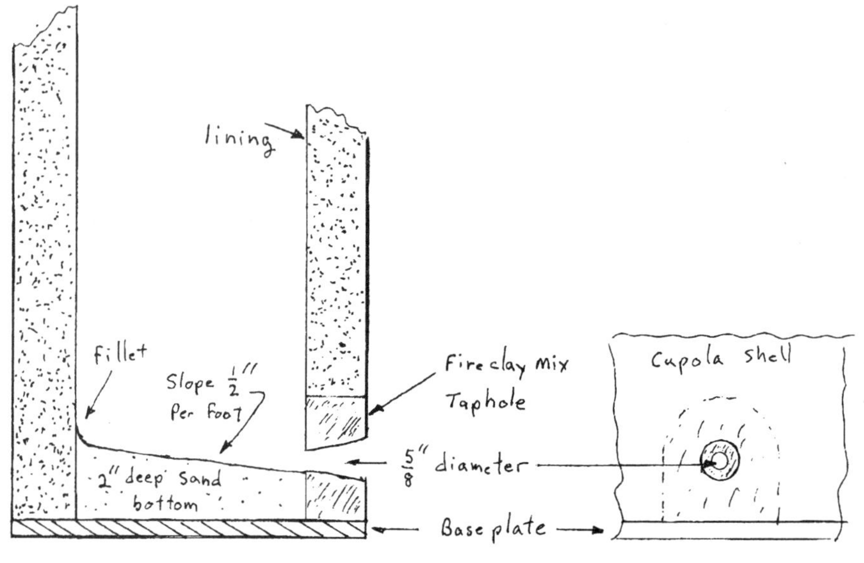

Sand Bottom   Taphole

The slag hole is made by opening the slag door and driving a knife through the refractory with a hammer. Cut an opening as shown in the photo.

Slag Door

FIRE THE LINING:

As soon as the lining is in spray the furnace with high temperature grill paint. Both the lining and the paint must cure at high temperature. Pour enough damp sand into the cupola to make a bed 2 inches deep. Open all the tuyeres. Soak several wads of newspaper in starter fluid and drop them down the shaft. Adding 1 cup of gasoline to 1 gallon of diesel can make a satisfactory starter fluid. Soak enough charcoal in starter fluid to make three layers in the cupola. Light the furnace by placing a lit propane torch through the taphole. When the charcoal is burning well, add more charcoal. This could take several hours, but you will want to have the whole cupola full of burning charcoal. You may want to put a blow drier or other small blower up to the tap hole to help the fire along. After about 5 hours, start adding coke and turn on a light blast. After another hour increase the blast and add more coke. Let it cook another hour then turn on full blast. The cupola will be extremely hot by now. After 30 minutes turn off the blast, open the tuyeres, and cover the top with one of the 15-inch circles you cut when making the wind belt. Let the fire burn out overnight. Drop the bottom and inspect the lining. Patch any holes. Your furnace is now finished.

Firing the lining is a slow process. If you get the lining too hot too fast, the outer surface of the lining will vitrify but the inside will still be wet. When you turn the heat up, steam will build up in the lining causing it to crack and slough off. Be patient and take your time.

Completed Cupola

III. OPERATION OF THE CUPOLA

SEQUENCE OF OPERATION:

1. Set Up Work Area
2. Weigh Charges
3. Make the Taphole
4. Make the Bottom
5. Make Slaghole
6. Lighting and Burning in the Bed
7. Charge
8. Soak or Preheat
9. Melting
10. Tapping
11. Dropping the Bottom
12. Cleaning the Cupola

Before the first melt you will have to mark a rod to the proper bed height. The intial bed height should be 18 inches above the tuyeres. Use a bod rod to measure down from the top of the cupola to the top of the coke bed, mark the handle. By using a bod rod you will always have the measuring rod available.

Because Cupolas are fast melters actual melting time will probably not exceed an hour. However preparation of the cupola for melting could exceed three or four hours. Each step should be carefully carried out. Once melting starts you will not have time to prepare charges, search for tapping bars, hammers or bod clay. Keep everything organized as shown in the site layout.

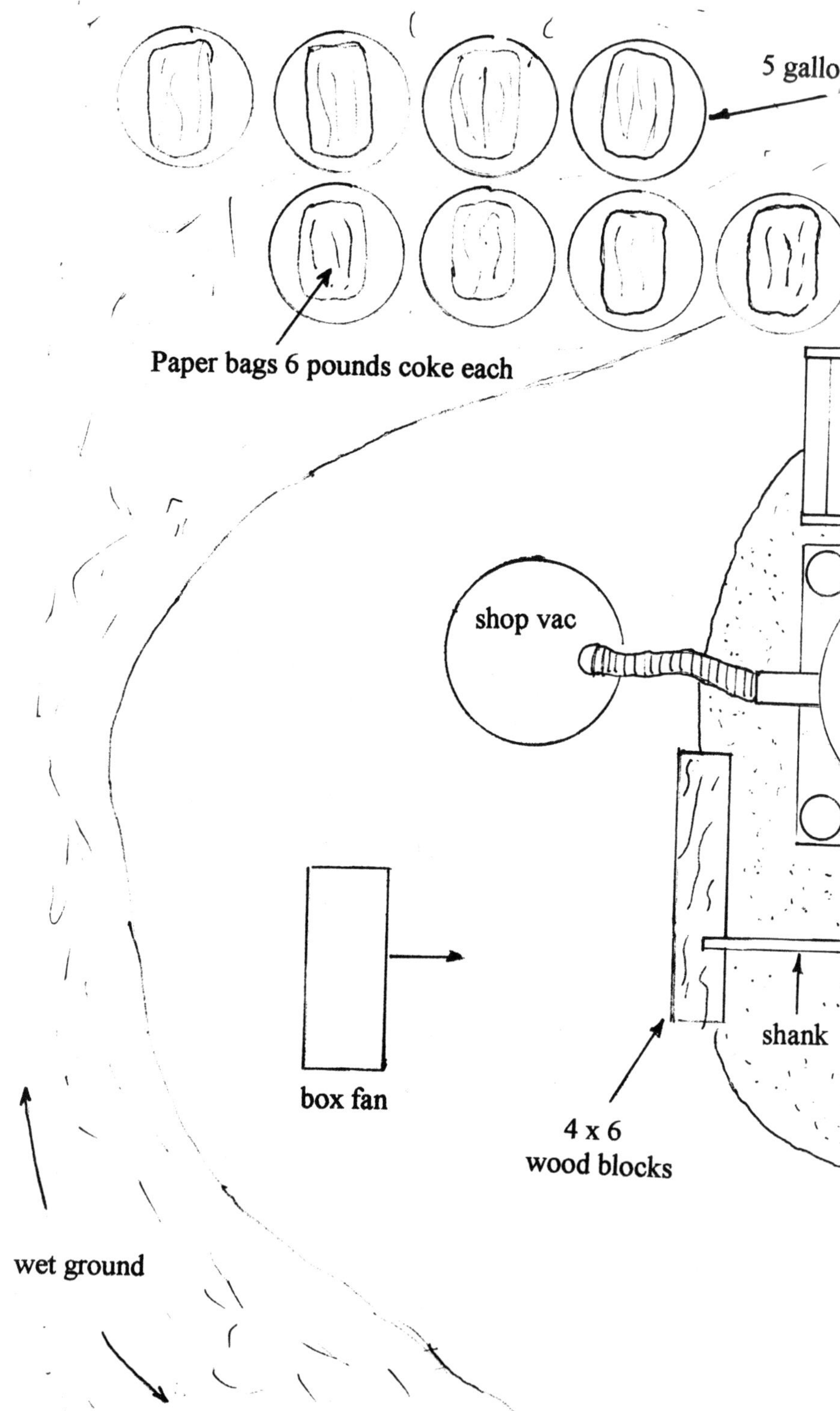

5 gallo
Paper bags 6 pounds coke each
shop vac
box fan
4 x 6
wood blocks
shank
wet ground

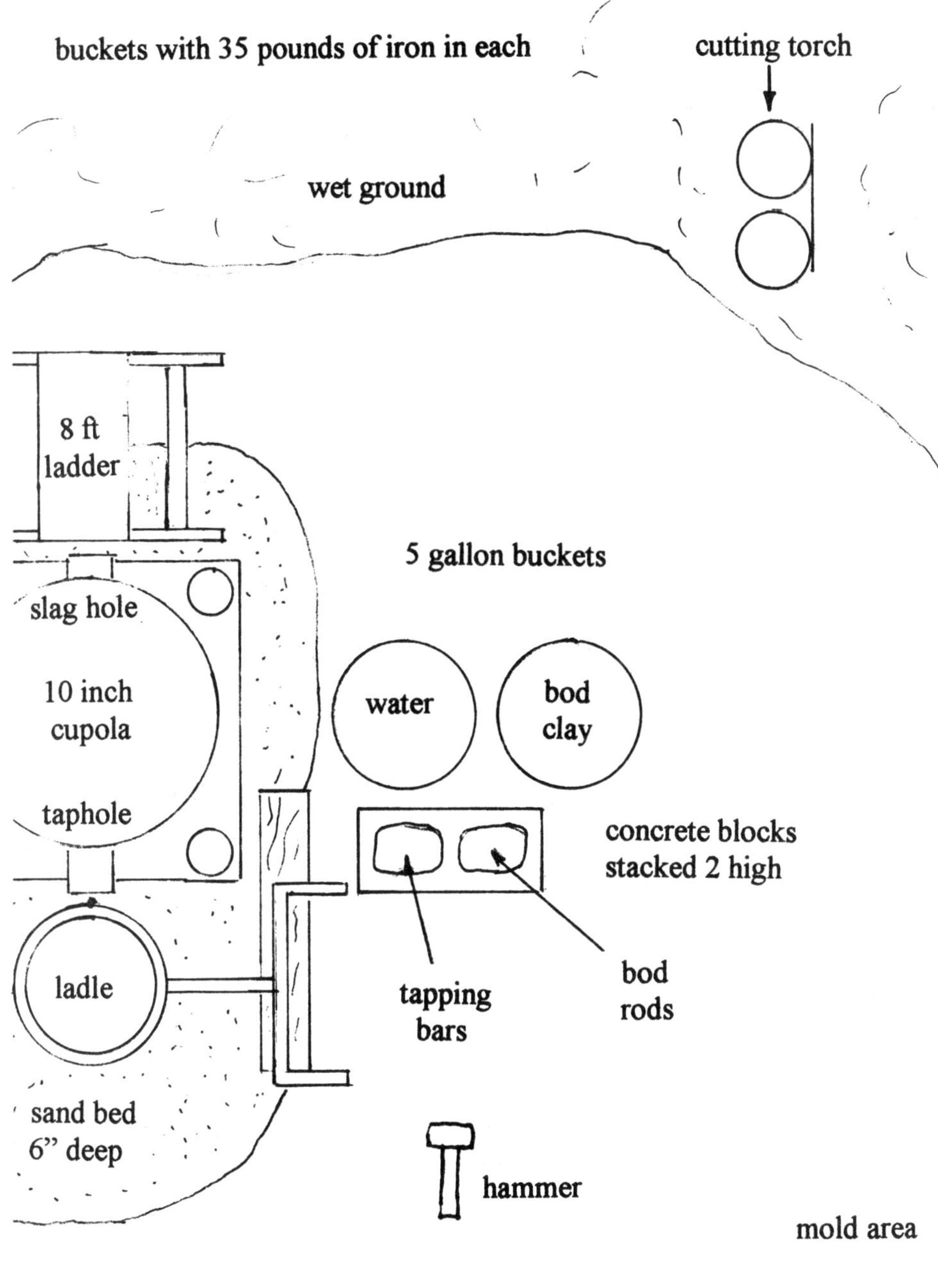

buckets with 35 pounds of iron in each
cutting torch
wet ground
8 ft ladder
slag hole
10 inch cupola
taphole
5 gallon buckets
water
bod clay
concrete blocks stacked 2 high
ladle
tapping bars
bod rods
sand bed 6" deep
hammer
mold area
SITE LAYOUT

1. SET UP THECUPOLA:

Set up the cupola on a firm foundation away from trees and houses. A firm foundation could be a concrete slab or any footing that will not allow the fully charged cupola to tip wobble or lean.

The area immediately around the cupola should be dry. When hot iron contacts a wet surface steam is instantly formed. This sends a shower of molten iron around the whole area. Keep your work area dry and free of combustible materials. However, because of flying sparks, wet down the area <u>outside the work area</u> with a hose.

A 6-inch deep bed of sand should be built up around the cupola. It should be sloped slightly towards the center so any molten iron spilled will not run out of the bed area. A sheet of ¼ inch plate should be cut to fit between the legs. Attach at least six feet of chain to the plate. This plate will catch the drop at the end of the melt. If the drop falls into the sand bed it becomes a real mess, this makes it difficult to sort out the coke and iron from the sand. Having the iron and coke drop on the plate saves much time sorting after the heat. Use the chain to drag the plate from under the cupola.

2. WEIGH CHARGES:

Proper preparation of the charges is critical for cupola operation. Coke must be screened to proper size. The ideal size is 1/12 the inside diameter of the cupola. This is .83-inch for the 10 in cupola. Since you will not find screens of this size you must use 2 screens. Choose screens with ½ inch squares, the other with 1-inch squares. Sift the coke through the small screen to separate and discard the fines (half-inch coke could be used in a 7" cupola). Then run the screened coke through the large screen to separate the large pieces of coke. Coke between ½ and 1 inch in diameter will give the best results. You can get good results with coke between ½ inch and 1½ inches in diameter. I purchase "nut coke" it is 1½ inches and smaller. Large coke can be crushed in a coke crusher as described by Stewart Marshall.

Once you have filled several 5-gallon buckets with screened coke, start weighing the charges. Ten paper grocery bags are each filled with 6 pounds of coke. Ten five-gallon buckets are filled with 35 pounds of iron each. The iron charge can be 10 pounds of large iron, and measure ½ x 3 x 4 inches. The remaining 25 pounds should be smaller, thinner iron.

When all the coke and iron charges are weighed, place one folded bag of coke in each iron bucket. Place the charge buckets next to the ladder as shown in the site layout.

3. MAKE THE TAPHOLE:

The fireclay mixture that was used for the lining can now be rammed into the void left around the taphole. Once the mixture is in place pierce the taphole from the inside surface of the cupola with a piece of rebar. Move to the front and smooth the hole with a wet finger to form a taper towards the spout. Next, line the spout with the fireclay mixture. Cut a trough in the clay to carry the iron.

4. MAKE THE BOTTOM:

The bottom is made from a mixture of ¾ sand and ¼ Georgia clay. Any clay should work, if you find a baseball field, that clay will work also. Bottom sand should have a slight bond, but not the bond that fireclay would give. The bottom will set fairly hard under the heat of the cupola. A bottom made with fireclay would setup much too hard for the bottom to drop. You would probably have to chisel it out.

The bottom of the cupola is rammed with the weaker sand and clay mixture. It must be uniformly rammed so no iron leaks through. The bottom should be 2 inches deep and sloped towards the taphole ½ inch per foot. Fillet the edges where it joins the lining. Pack the joint between the top and bottom cupola shells with the mixture used on the bottom.

5. MAKE THE SLAG HOLE:

Build a mound of the fireclay mixture upon the back door and pierce it with a piece of rebar similar to the taphole. The door is now locked in place and the edges are carefully packed with the fireclay mixture to ensure no iron leaks develop around the door.

6. LIGHTING AND BURNING IN THE BED:

Open the tuyeres and light the coke fire using paper, charcoal and starting fluid. Place a small blower at the taphole to help the bed burn. Add coke, then wait for each charge to fully ignite before adding the next. Continue this process until the height of the burning bed is over the tuyeres. When the height of the burning bed is over the tuyeres, close them. Now use the exhaust from a shop-vac and apply a light blast. Remove the filter from the vacuum to get the maximum airflow. Frequently check the bed with the bod rod, marked earlier. When the burning bed is 18 inches above the tuyeres (calculated in chapter 1, it may be slightly higher, but no lower) add an additional 2 to 3 inches of coke. Apply full blast to preheat the furnace and burn the bed back to 18 inches above the tuyeres. The cupola will be very hot at this point. Now cut off the blast and open the tuyeres.

7. CHARGE:

Charge the cupola with iron and coke (iron first) in layers until the cupola is full. Usually 3 charges will fit.

8. SOAKING:

Let the cupola sit with no blast, tuyeres, slaghole and tap hole open for 45 minutes. This is to allow the charges to preheat. Rehearse a tapping and bodding operation during the preheat phase. Carry the empty ladle to each mold to ensure you can easily reach each mold and practice a pour. This rehearsal is important. You don't want to find out you can't reach a mold when you have 55 pounds of white hot iron in your hands.

9. MELTING:

Apply a 3-inch blast after the soaking period. During the next few minutes the charges will begin to settle and the indicated pressure may increase. The final blast pressure should be between 4 and 5 inches.

After 8 to 10 minutes iron should be visible at the taphole. If it is seen before this, the bed is too low. Decrease the blast. If more than 12 minutes pass the bed is too high. Increase the blast to burn away the excess coke.

The first iron will be cold. It may start to freeze in the spout and taphole. Work quickly to knock it out of the way with a piece of rebar. Be careful not to let the rebar freeze in the tap hole or you will have to cut it out with a torch. You must work very quickly here! Once the stream of iron is steady, bod it off. Wait 2 to 3 minutes to let the iron build up in the well and heat the sand bottom. Tap the cupola. This first iron will be cold and should be poured into an ingot mold. Charge the cupola with iron and coke and prepare for the next tap.

10. TAPPING THE CUPOLA:

Tap the cupola by picking out the bod with a tapping bar. A tapping bar is a 3-foot section of rebar with a chisel point on one end and a T handle on the other. The iron may start flowing with a little poking or it may require breaking through a frozen iron plug. Hammering the tapping bar through with a 2-pound hammer can usually break plugs. Be quick to extract the tapping bar so the bar does not freeze in the taphole. At the first sign of slag close the taphole with a bod. Frozen slag in the taphole makes tapping very difficult. After tapping the cupola the taphole must be closed with a cone shaped plug called a bod. The bod is roughly the size of a golf ball and is made of the same sand and fireclay mixture that was used for the spout. This mixture sets up fairly hard and might be better is some sawdust were added. The sawdust would make the bod softer and easier to pick out. A bod rod is made of a 3-foot section of rebar with a T handle on one end, and a 1 ½-inch disc welded to the other. I cut my disc from a rod, however ¼ inch plate also works well. The disc end of the bod rod is placed in a bucket of water, then a wad of clay is formed into a cone over the wet tip. This rod is placed, standing up, in the concrete blocks as to be ready when needed. It is a good idea to keep at least 2 bod rods and tapping bars ready incase a bod falls off, or the tapping bar becomes coated with iron.

For larger taps, up to 55 pounds, the slaghole is bodded off when iron reaches it. Wait 4 to 5 minutes, then tap the cupola.

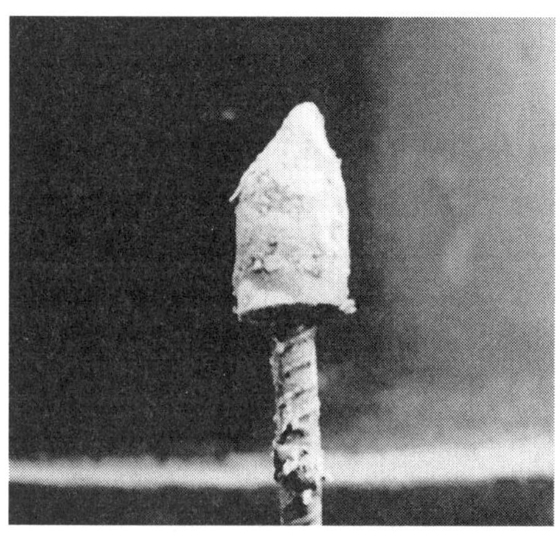

Clay Bod

11. DROPPING THE BOTTOM:

Tap all the liquid iron remaining in the cupola at he end of the melt. Turn off the blast and open the door latch. A properly constructed sand bottom will support the mass of unburned coke, but will rupture readily when picked with the tapping bar. <u>The cupola's contents will be extremely hot.</u> It is only a matter of a few seconds before you will have to step away from the drop. After you have cooled down, prod the bottom again to be sure nothing is lodged in the cupola. When you are confident the contents have fully dropped, spray the drop with a hose. Keep a safe distance from the drop when spraying it with water. Steam will be violently liberated and you could easily be burned.

11. CLEANING THE CUPOLA

The next day, when the cupola has cooled, it may be cleaned and patched (if any refractory has cracked and sloughed off). A pointed steel bar and hammer may be used here. A pneumatic chisel also helps speed the clean up. Break out the slag so that future-descending charges will not hang up; however don't

Cleaning the cupola with a pneumatic chisel

remove the smooth glaze that has formed on the lining. Knock out the tap hole with a 1-inch steel rod and the 2-pound hammer. Be sure the lining is smooth and free of large lumps.

After several heats I prefer an alternative to dropping the bottom. At the end of the heat all of the iron is tapped out and blast is turned off. The cupola is allowed to cool overnight. The next day the cupola is inverted and the bottom is knocked out with a sledgehammer. The cleanup is finished with a pointed bar and hammer.

IV. AIR SUPPLY:

The balance of air and coke is critical for proper cupola operation. Cupola blowers should be sized at 2.5 CFM per square inch of sectional area between the tuyeres. Besides CFM, static pressure is also important. Static pressure is the pressure a blower develops if it is connected to a sealed container. As static pressure in the cupola windbelt increases, the CFM decreases. Static pressure is related to the physical size of the coke and iron pieces that make up the charges, how tightly packed the charges are, and the height of the stack. In other words, resistance to airflow increases the static pressure. Fans are rated by CFM and by static pressure. It does no good to have the proper CFM if there is not enough pressure to push it through the cupola's coke bed and stack.

Pressure blowers supply the combustion air for cupolas. Pressure blowers come in two types, positive displacement, and centrifugal fans. Roots type blowers are the positive displacement type. With each revolution of the blower lobes, a known quantity of air is compressed. Roots blowers are capable of pressures over 25 PSI. Maximum rated output of common roots blowers is shown in the chart below. Roots blowers are commonly used on mosquito spraying equipment. Many city and county pest control departments routinely junk such blowers. Dresser Industries manufactures roots blowers and has distributors in most states. You may be able to find a used blower through one of their pump dealers.

Centrifugal fans are more commonly used on cupolas. The six types of centrifugal fans are classified by blade type. In order of decreasing efficiency they are as follows: 1) airfoil blades 92% efficient, 2) backward curved blades 85%, 3) backward inclined blades 78%, 4) radial tipped blades 70% 5) forward curved blades 65% and 6) radial blades 50%. We are only concerned with radial blade high-pressure blowers A high-pressure blower's output is measured in ounces or in inches water column. 1 ounce of pressure will raise a column of water 1.73 inches. Water column is measured by a manometer.

Frame Size	Speed RPM	Inlet Vac. Inches Hg. (kPa)	Temp. Rise Fahr. Deg. (C°)	Press. Rise PSI (kPa)
22	5275	15 (50)	225 (125)	12 (82)
24	5275	15 (50)	210 (117)	7 (47)
32	3600	15 (50)	225 (125)	15 (101)
33	3600	15 (50)	225 (125)	12 (82)
36	3600	15 (50)	225 (125)	7 (47)
42	3600	15 (50)	240 (133)	15 (101)
45	3600	15 (50)	225 (125)	10 (68)
47	3600	15 (50)	225 (125)	7 (47)
53	2850	15 (50)	225 (125)	15 (101)
56	2850	15 (50)	225 (125)	10 (68)
59	2850	15 (50)	225 (125)	7 (47)
65	2350	16 (53)	250 (139)	15 (101)
68	2350	16 (53)	240 (133)	12 (82)
615	2350	12 (40)	130 (72)	6 (40)
76	2050	16 (53)	250 (139)	15 (101)
711	2050	16 (53)	210 (117)	10 (68)
718	2050	12 (40)	130 (72)	6 (40)

Operating Characteristics of Small Roots Blowers

A manometer is a glass or plastic tube in the shape of a "U" that is partially filled with colored water. One side of the tube is connected to the windbelt; the other side is left open to the air. The pressure in the windbelt forces the one side of the water column down and the other side up. The difference in height of the two water columns is measured in inches with a ruler.

SELECTION OF CUPOLA BLOWERS:

Cupolas will normally melt 4000 pounds of iron for every 1000 cfm of air supplied. The 10 inch cupola is melting 330 pounds / hour at 90 cfm. It is operating at approximately 92% of the above general rule. Usually cupola blowers are sized at 2.5 cfm per square inch of sectional area between the tuyeres. The recommend blower for the 10-inch cupola is:

Area of the cupola = 78.5 inches^2

2.5 cfm X 78.5 = 196 cfm, approximately 200 cfm

Blowers are frequently sized up 10% to make up for leaks in the system and variations in temperature. However they are often run at 80% to 90% of the calculated value.

Operating pressure is related to system resistance. Recommended fan pressures are shown on the graph. The cupola wind belt will probably not reach these pressures; they are only recommended blower capacity.

Blower Sizes for Cupola Operation

Inside Diameter in inches	Area inches^2	Actual cfm	Recommended Blower size	
			cfm	discharge pressure oz.
10	78.5	165	216	8
18	254	510	700	16
23	415	830	1140	20

The graph below illustrates suggested blower pressure capacity for sectional area between the tuyeres. To use the graph find the section area on the bottom, draw a line straight up until you

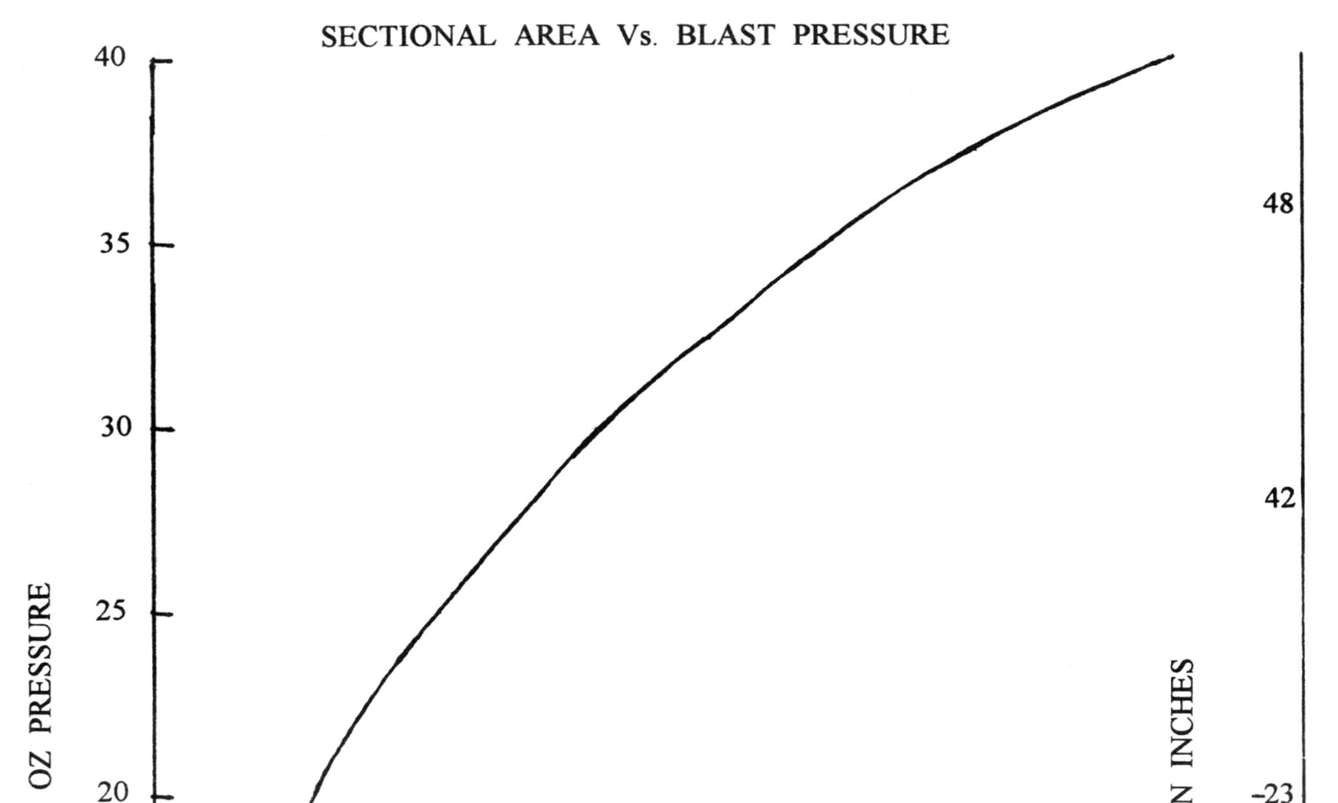

SECTIONAL AREA Vs. BLAST PRESSURE
OZ PRESSURE
40
35
30
25
20
IN INCHES
48
42
-23

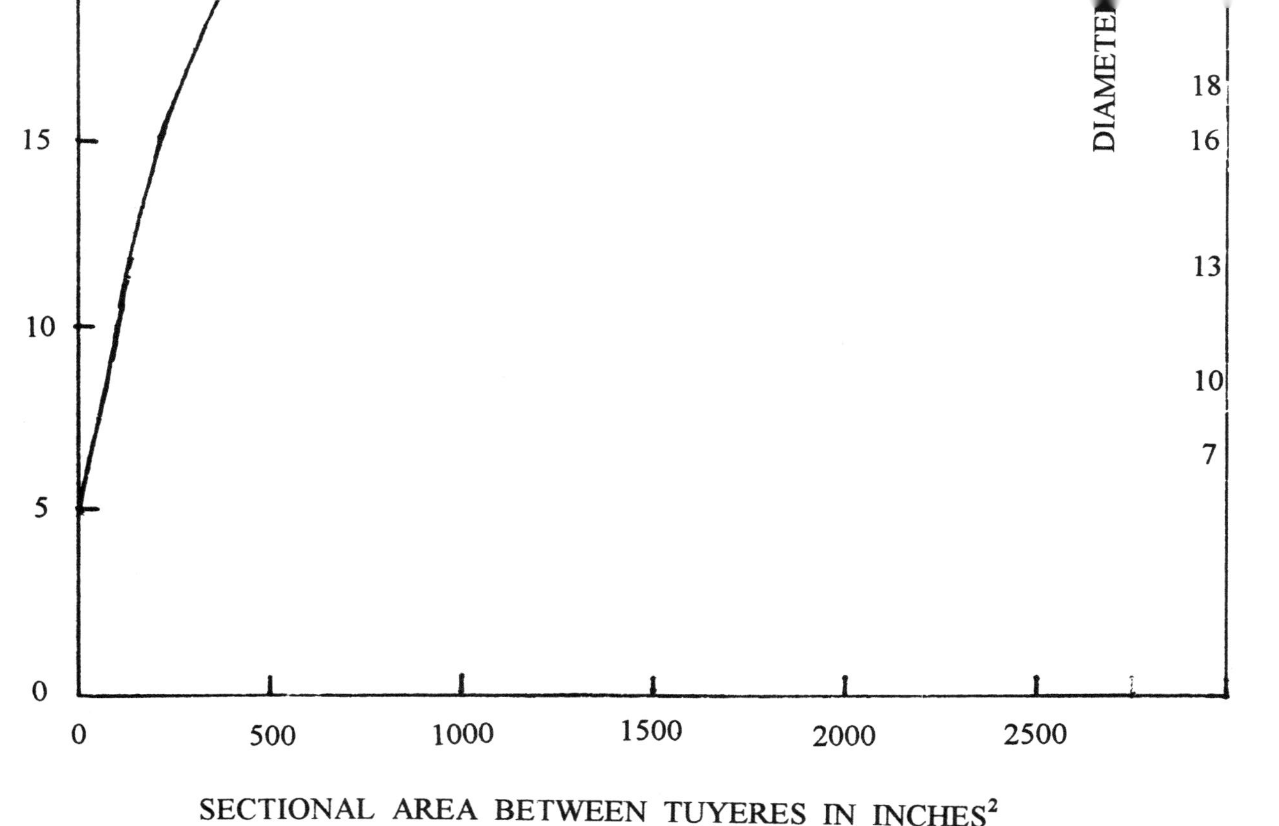

DIAMETER
18
16
13
10
7
15
10
5
0
0
500
1000
1500
2000
2500
SECTIONAL AREA BETWEEN TUYERES IN INCHES²

ROOTS TYPE BLOWER
PERFORMANCE 2510J WHISPAIR MAX
MAXIMUM PRESSURE RISE = 4 PSI
BASED ON INLET AIR AT 14.7 PSI & 68° F

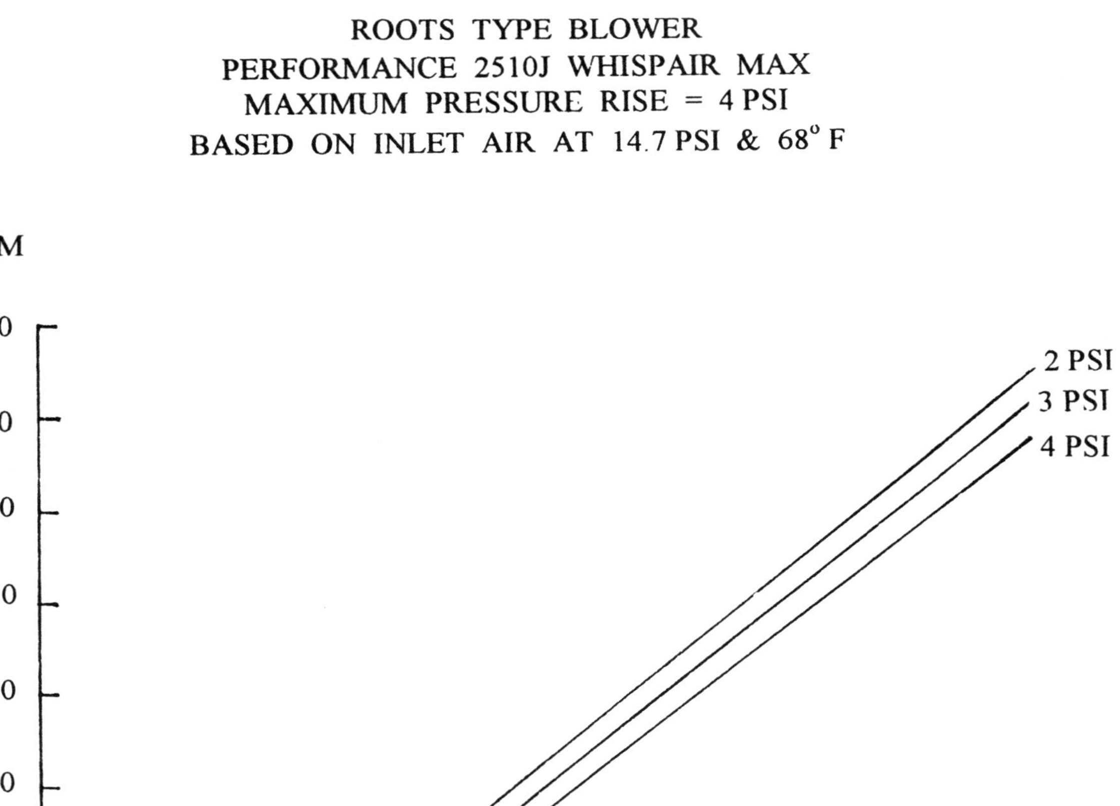

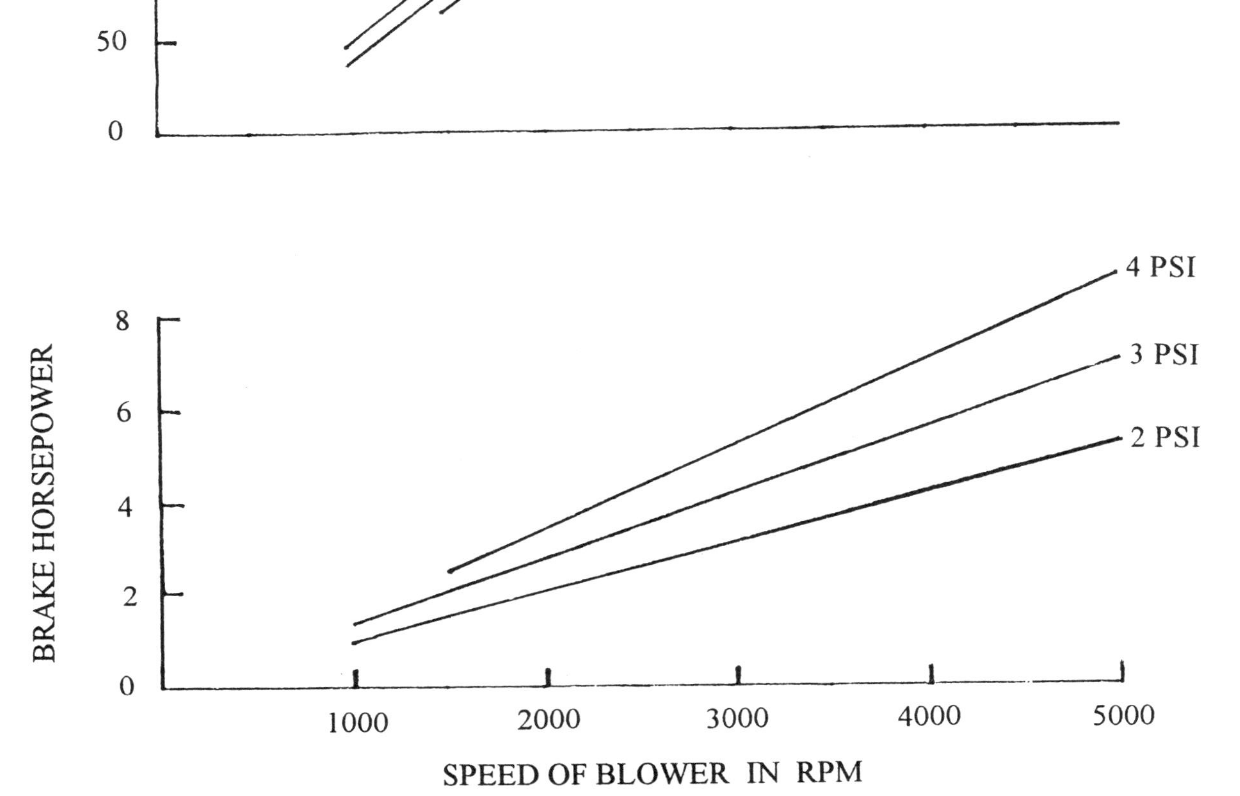

8
6
4
2
0
BRAKE HORSEPOWER
1000
2000
3000
4000
5000
SPEED OF BLOWER IN RPM
4 PSI
3 PSI
2 PSI
50
0

intersect the curve, then draw a horizontal straight line to the left. The pressure will be indicated on the left axis.

A salvaged Roots type blower with input muffler

V. DESIGNING CENTRIFUGAL FANS:

Design of propeller type fans is fairly straightforward. A satisfactory prototype may be built from the equations for propeller type fans. Formulas for centrifugal type fans are not as accurate. Approximate dimensions may be calculated, however actual performance must be measured after assembly. The fan speed or dimensions may be adjusted to bring the fan into compliance with the original design goals.

After commercial fans are built, they are tested. The performance is plotted on a graph called a fan output curve, or performance graph. Fan curves are very useful tools to the fan designer. If you know the actual dimensions and rpm of the fan plotted on the fan curve you can use formulas called the fan laws to convert fan performance of a known fan to meet new requirements. Once you build and test one fan you can size it up or down and, or change the speed and have predictable results, provided <u>all dimensions remain geometrically proportional.</u> This includes such factors as number of blades, front and rear plates, inlet and outlet area. In other words, the new fan must be an exact scaled copy of the original.

DESIGNING WITHOUT PERFORMANCE CURVES:

Fan output is related to wheel diameter, width, speed, number of blades, inlet and outlet areas. Once you know the cfm and static pressure fan must produce, you may start with the formula below.

The minimum wheel diameter:

Outside diameter inches = $(15500/\text{rpm})\sqrt{SP}$

SP = static pressure

The constant 15500 may increase for less efficient radial blade designs, or decrease for the more efficient designs.

The minimum inside diameter, or inlet size:

$$\text{inside diameter} = 11\sqrt[3]{(\text{cfm/rpm})}$$

width of the wheel = (550 cfm)/(D x tipspeed) ft/min

D = diameter in inches

Tip speed is the speed of the outside edge of the wheel.

$$\text{Tip speed} = (\pi D_{inches}\text{rpm})/12$$

Fan efficiency is assumed to be 50%

Horsepower required by fan:

Hp = (cfm x inches wc) / 3178   wc = water column

DESIGNING A WHEEL FOR THE 10 INCH CUPOLA:

Parameters:

cfm = 220

static pressure = 8 oz

convert oz to inches wc: (8 oz) x 1.73 inches/oz = 13.84 inches

Since motors run approximately 3500 rpm we will start here. This may or may not provide us with a workable design.

$$\text{outside diameter} = (15500/3500)\sqrt{13.84} = 16.475 \text{ inches}$$

tip speed = $(\pi \times 16.475 \times 3500) / 12 = 15096$ ft/min

Width of wheel = $(550 \times 220) / (16.475 \times 15096) = .487$ inches

The .487inch wide blade will be difficult to assemble, lets increase the rpm.

rpm = 5700

od = $(15500/5700)\sqrt{13.84} = 10.1$ inches

rounding to 10 inches, tip speed is 14,922 ft/min

width of wheel = $(550 \times 220) / (10 \times 14,922) = .81$ inches

.81 inches will be easier to manufacture and a 10-inch wheel requires less horsepower, therefore we will use this wheel.

Calculate the inlet diameter:

Minimum inlet diameter = $11 \sqrt[3]{(220 / 5700)} = 3.72$ inches

Calculate the horsepower required by the fan:

hp = $(220 \times 13.84) / 3178 = .96$ hp

MECHANICAL STRESSES DEVELOPED IN CENTRIFUGAL FANS:

Balance of centrifugal fans is critical. If the wheel of our 10-inch fan is out of balance by the weight of a sheet of paper , a rotating force of 30 pounds will be developed. The vibration caused by such an imbalance would soon cause the fan to fly apart. Fortunately balancing a fan is not difficult if you build a set of balancing ways. Construction of balancing ways will be described later.

The centrifugal force formula:

Force = (W x R x rpm^2) / 2933

W = weight in pounds

R = distance in feet from the center of the blade to the center of the fan wheel.

Lets calculate the centrifugal force on one of the fan blades of the 10-inch wheel. If the blade is made from 16-gage steel it would weigh approximately 1.45 oz.

Convert ounces to pounds: 1.45oz /16oz / pound = .091 pound

Force = $((.091_{pound})$ x $(.348_{foot})$ x $(5700^2))$ / 2933 = 350 $_{pounds}$

If the fan were to come apart the blades would have a force of 350 pounds and a speed of 170 mph. As you can see the forces generated in this small fan are considerable. Although fan construction is not difficult it does require precision work and precision balancing. A poorly constructed fan will at best explode and wreck the afternoon's melt. It could also dismember or kill someone. A properly constructed fan is a precision tool. Treat it as such and you will have no problems.

SCROLL HOUSING:

The individual streams of air leaving the blade tips are collected and formed into a single air stream that is discharged through the housing outlet. Three circular sections form the spiral. The radii of these sections are .5D, .7D, and .9D. The radii are off set from the center of the wheel by .1D.

Calculate the scroll radii for the 10-inch wheel :

$R_1 = .5 \times 10$ inches $= 5$ inches

$R_2 = .7 \times 10$ inches $= 7$ inches

$R_3 = .9 \times 10$ inches $= 9$ inches

Calculate the offset:

Offset $= .1 \times 10 = 1$ inch

Discharge opening is calculated by: $2(D \times W) / 3$

Discharge area $= 2(10 \times .82) / 3 = 5.46$ inches^2

Cut off point: the cutoff point is where the scroll is closest to the wheel. The closer the cutoff point is to the wheel the noisier the fan will be. A good figure is 5% of the wheel diameter.

Cutoff point $= .05D$

Cutoff point $= .05 \times 10$ inches $= .5$ inches

DESIGNING FROM AN EXISTING FAN:

After you have designed and tested a fan, or have a known design with a fan performance graph, you may calculate the performance of another geometrically similar fan, i.e. "an exact scaled copy." To convert fan performance, use the following rules:

To convert the performance of a fan from one speed to another:

cfm varies directly with speed:

$$(cfm_2 / cfm_1) = (rpm_2 / rpm_1)$$

Calculate the new cfm if the speed is increased:

$cfm_1 = 200$

$rpm_1 = 1800$

$rpm_2 = 3600$

$$cfm_2 = (cfm_1 \ rpm_2) / rpm_1$$

$$cfm_2 = (200 \times 3600) / 1800$$

$$cfm_2 = 400$$

Pressure varies as the square of the speed.

$$(SP_2 / SP_1) = (rpm_2 / rpm_1)^2$$

$$SP_2 = SP_1 \ (rpm_2 / rpm_1)^2$$

$SP_1 = 3$ inches wc

$rpm_1 = 1800$

$rpm_2 = 3600$

$$SP_2 = 3 \text{ inches } (3600 / 1800)^2$$

$$SP_2 = 12 \text{ inches}$$

Calculate new rpm if SP is known:

$$rpm_2 = rpm_1 \sqrt{SP_2 / SP_1}$$

$rpm_1 = 3600$

$SP_1 = 5.2$

$SP_2 = 13.84$

$$rpm_2 = 3600\sqrt{13.8 / 5.2}$$

$$rpm_2 = 5865$$

Horsepower varies as the cube of the speed:

$$(hp_2 / hp_1) = (rpm_2 / rpm_1)^3$$

$$hp_2 = hp_1 (rpm_2 / rpm_1)^3$$

$hp_1 = .125$

$rpm_1 = 1800$

$rpm_2 = 3600$

$$hp_2 = .125(3600 / 1800)^3$$

$$hp_2 = 1$$

Changing the wheel diameter:

cfm varies as the cube of the size.

$$(cfm_2 / cfm_1) = (D_2 / D_1)^3$$

$$cfm_2 = cfm_1 (D_2 / D_1)^3$$

$$cfm_1 = 200$$

$$D_1 = 10 \text{ inches}$$

$$D_2 = 14 \text{ inches}$$

$$cfm_2 = 200cfm (14 \text{ inches} / 10 \text{ inches})^3$$

$$cfm_2 = 549$$

Static pressure varies as the square of the diameter.

$$(SP_2 / SP_1) = (D_2 / D_1)^2$$

$$SP_2 = SP_1 (D_2 / D_1)^2$$

$$D_1 = 10 \text{inches}$$

$$D_2 = 14 \text{ inches}$$

$$SP_1 = 5.2 \text{ inches}$$

$$SP_2 = 5.2 \text{ inches}(14 \text{inches} / 10 \text{ inches})^2$$

$$SP_2 = 10.2 \text{ inches}$$

Horsepower varies as the fifth power of the diameter.

$$(hp_2 / hp_1) = (D_2 / D_1)^5$$

$$hp_2 = hp_1 (D_2 / D_1)^5$$

$D_1 = 10$ inches

$D_2 = 20$ inches

$hp_1 = \frac{1}{2}$

$$hp_2 = \frac{1}{2} \, hp\,(20 \text{ inches} / 10 \text{ inches})^5$$

$$hp_2 = 16$$

In order to design the best fan for your application you must know which parameter is the most important, static pressure, cfm or horsepower input. The following conclusions may be helpful:

Increasing wheel speed increases the SP faster than the cfm because SP is a squared factor while cfm is directly proportional to the wheel speed.

Increasing the wheel diameter increases the cfm faster than the SP because the cfm varies as the cube of the diameter, and SP varies as the square of the diameter.

Power consumption increases very rapidly for an increase in diameter because the horsepower varies as the fifth power of the diameter. For example, if a 1 horsepower 12-inch wheel is increased to 24 inches, the new power required will be 32 horsepower.

CALCULATING PULLEY SIZES:

Obtaining the correct fan speed from the motor speed is calculated from the ratio:

$$RPM_1 / P_2 = RPM_2 / P_1$$

RPM_1 = Motor speed P_1 = Motor Pulley Diameter

RPM_2 = Fan speed P_2 = Fan Pulley Diameter

The formula may be easily manipulated to solve for either pulley size if one pulley size and the speeds of the motor and fan are known.

$$P_1 = (P_2 \times RPM_2) / RPM_1$$

$$P_2 = (P_1 \times RPM_1) / RPM_2$$

Calculate the pulley diameter for a 3600-RPM motor if the desired fan speed is 5700 RPM and the fan pulley is 2 inches in diameter:

$$P_1 = (P_2 \times RPM_2) / RPM_1$$

$$P_1 = (2 \times 5700) / 3600$$

$$P_1 = 3.167 \text{ inches diameter}$$

VI. CONSTRUCTION OF CENTRIFUGAL FANS:

Fan construction is at least a full day's work, and most likely a two-day project. Centrifugal fans may be built of sheet metal, or steel plate. The 10-inch wheel may be constructed of 20-gauge steel, however a sturdier metal such as 16 gauge should be used for the scroll housing. The thinner metal, when used for the scroll housing, tends to collapse toward the wheel during operation. For the larger or direct drive fans, a scroll housing back plate might be built of ¼ inch plate, drilled to receive a C-faced motor, such as the motors used on pumps.

Vanes may be held on the wheel by steel pop rivets. As we have seen earlier, the forces generated in a fan are considerable, ruling out aluminum rivets. Larger wheels may have welded vanes.

Depending on the thickness of the material chosen, the wheel may be cut out with tin snips, a jig saw, band saw, or torch. With much effort, 20-gauge material may be cut with compound snips; 16 gauge may be cut with a jigsaw, or preferably a band saw; ¼ inch plate requires a torch. My first several wheels were cut with hand tools, however a rotary table and a mill make the whole fan project much faster and easier. Hand construction methods are described here.

Using dividers, draw two wheels (the front and rear plates) on the sheetmetal. Carefully cut the wheels out without bending or distorting them. You may cut the wheels a little larger than the scribed line and clean them up by grinding to scribed mark. Draw a horizontal line through the center of one circle. Using a protractor, carefully scribe the locations of the blades. To find degrees between the blades, divide 360° by the number of blades. For the twelve blades used on the 10-inch wheel there are 30° between each blade. Scribe two circles to locate the rivets for the blades. On the remaining wheel draw the circle for the inlet area.

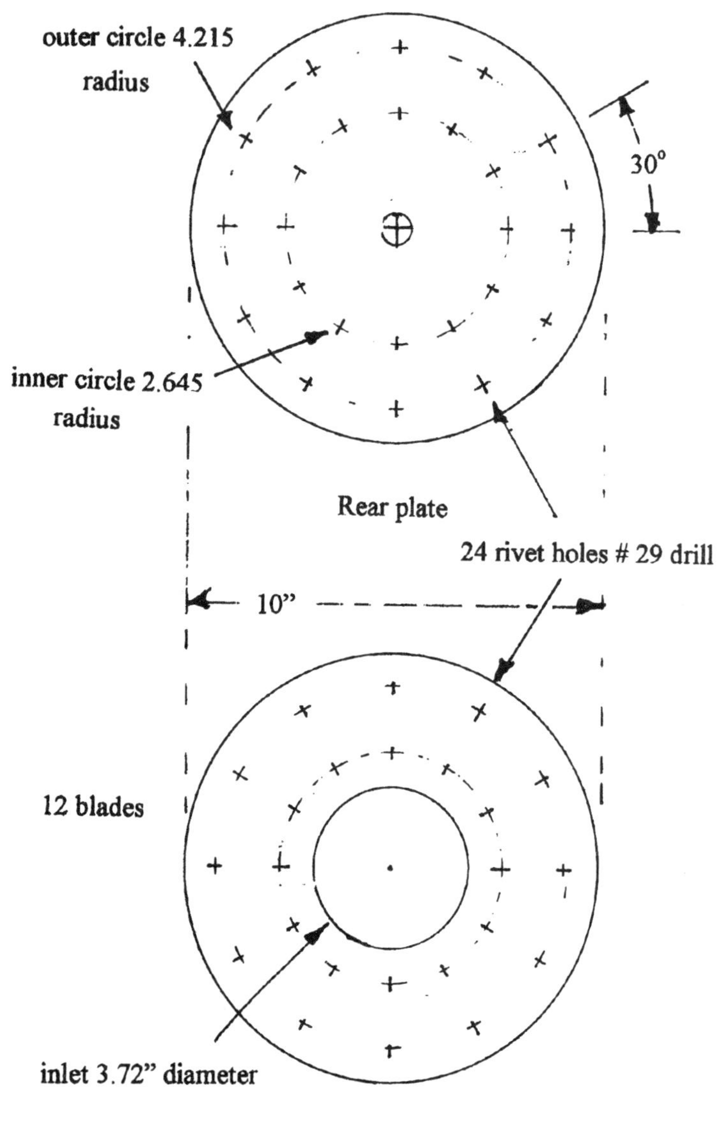

Plate layout for 10" cupola fan

Center punch the hole in the center of the wheels and all the rivets holes on one of the back plates. Locate and punch the holes for the mounting hub. All of these holes must be

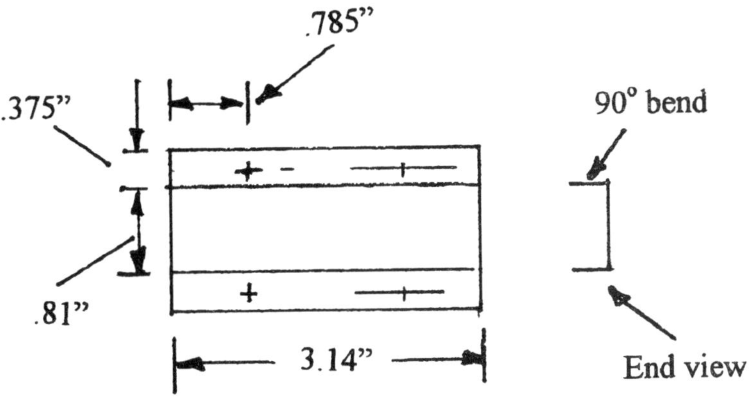

Fan Vane

accurately located, or the wheel will be impossible to balance. Clamp the wheels together and drill two rivet holes on both halves of the wheel. Rivet the wheels together and finish drilling the rivet holes. Remove the rivets, drill the center hole the size of the shaft you are using (motor shaft or mandrill). This is best accomplished by screwing the wheel to a section of plywood and step drilling, increasing the drill diameter by no more than 1/8 inch each pass. Sheet metal has the tendency to climb up the bit, so securely attach the wheel to the plywood.

If you are lucky enough to have a trepanning tool and drill press, the inlet hole is easily drilled. Otherwise you will have to cut it with snips and grind it to final size, or drill a series of small holes just inside of the scribed line. The inner circle is removed and discarded. The hole is ground and /or filed to final size.

The vanes are cut out, although the plan calls for .81 inch, I increased the width to 1.1 inches so that I might use the fan on a 12-inch cupola. This worked out well as the fan can be throttled down for the smaller cupola. The mounting tabs are scribed and bent. Make the bends by clamping in a vise and using a hammer to pound the bend flat. If you have a triple beam balance, now is the time to weigh and grind the vanes until their weight is within 1/10 of a gram of each other. This cuts your balancing time down to a minimum. If you do not have a balance, grind them until they are as identical in size as possible.

The centerline of each mounting tab is marked. The location of one of the rivet holes is drilled on each mounting tab. The second set of holes will be drilled after the vane has been riveted in place with one set of rivets.

Once the vanes are riveted in place install the hub. A hub may be turned on the lathe, salvaged from an old fan, or made from an old pulley. Once the hub is installed, balance the wheel on a set of balancing ways. The balancing ways require the wheel be mounted on a polished shaft to work properly. Drill rod is preferred. The slightest surface imperfection will ruin the accuracy of the balancing ways. This means galvanized shafting is useless for balancing.

Balance the wheel by removing material from the heavy side of the wheel. When placed on a set of balancing ways the wheel will roll until the heavy side is down. Test the wheel and reverse the wheel on the ways, if they are truly level you will get the same result (heavy spot) each time. If the wheel has been well constructed, it may take only light sanding of the heavy edge with a belt sander. Otherwise grinding and drilling the heavy side should work. Continue testing and trimming until the wheel will rest in any position on the ways without moving. Balancing can be a time consuming process. Care used in building the wheel will minimize the amount of time spent balancing.

Rear Plate with Vanes Attached

BEARINGS:

Pillow block bearings are poured from aluminum. Bore the pillow blocks by using an angle plate on the faceplate of the lathe. Number 6203 ball bearings are fitted into the castings. These bearings are rated to run at 17,000 rpm, so 5800 rpm used in the fan should be acceptable.

Purchasing the pillow blocks will be easier; however 6203 bearings are so cheap, $2.95, I decided to make my pillow blocks.

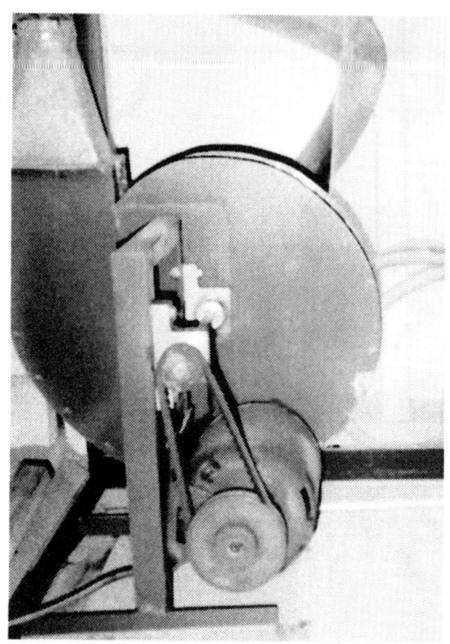

Shop made fan joined with sheet metal screws

BALANCING WAYS:

Balancing ways must be nearly frictionless and absolutely level. A good set of balancing ways may be easily built using precision steel rules for the frictionless edge. A frame is built to support and level the rules. The rules are attached with cap screws. The screws should be snug but not overly tight. This may distort the rules giving in accurate results. Because the rules do not have holes drilled in them, but rest on top of the cap screw threads, shims may be needed under the bottom edge of each the washers to maintain a flat surface for the cap screws, and prevent distortion of the rules.

The frame is welded up from 1-inch angle. My angle was salvaged from an old bed frame. The hinge was made from 1/8 by 1-inch strap. All of the bolts are ¼ - 20. The base of the balancing ways is constructed of a 6 inch by 10-inch section of Masonite. The distance between the ways is kept small to minimize the inertia of the shaft. Later, substituting a wider section of Masonite will allow you to balance wider wheels and motor rotors.

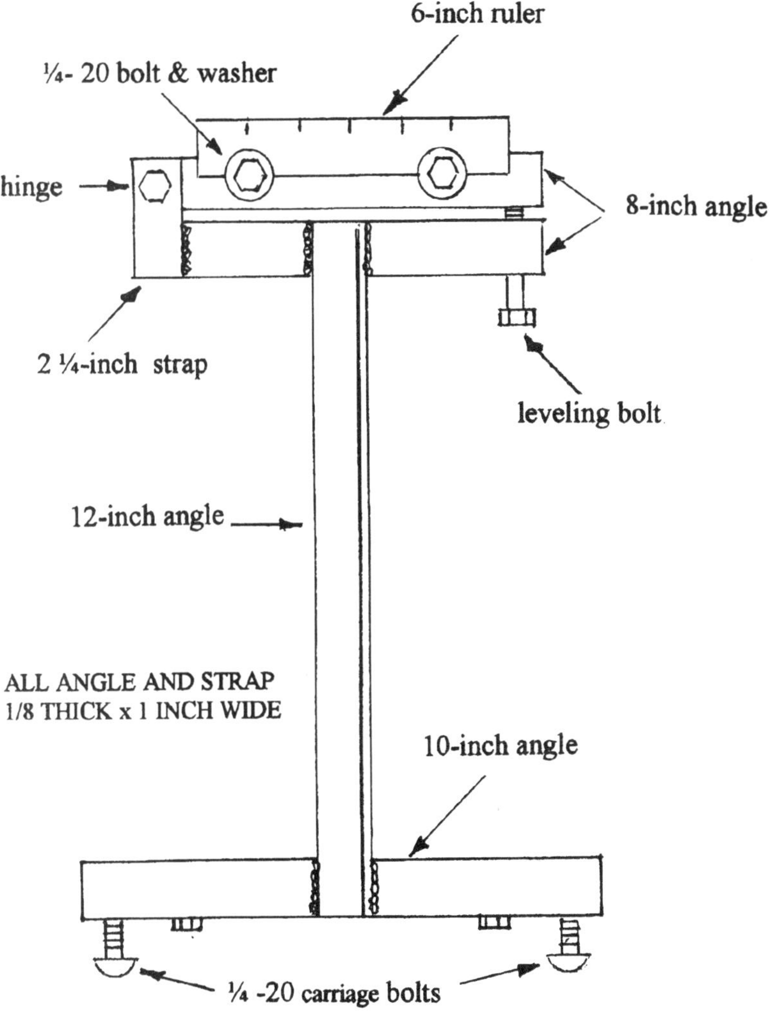

Balancing Ways Side View

Set the ways up on a level surface and adjust the feet until there is no wobble. After each of the ways is leveled, check between both of the ways. If your level is not long enough to reach across the distance between the ways, use a piece of polished shaft or drill rod to span the distance. Recheck the ways until they are absolutely level.

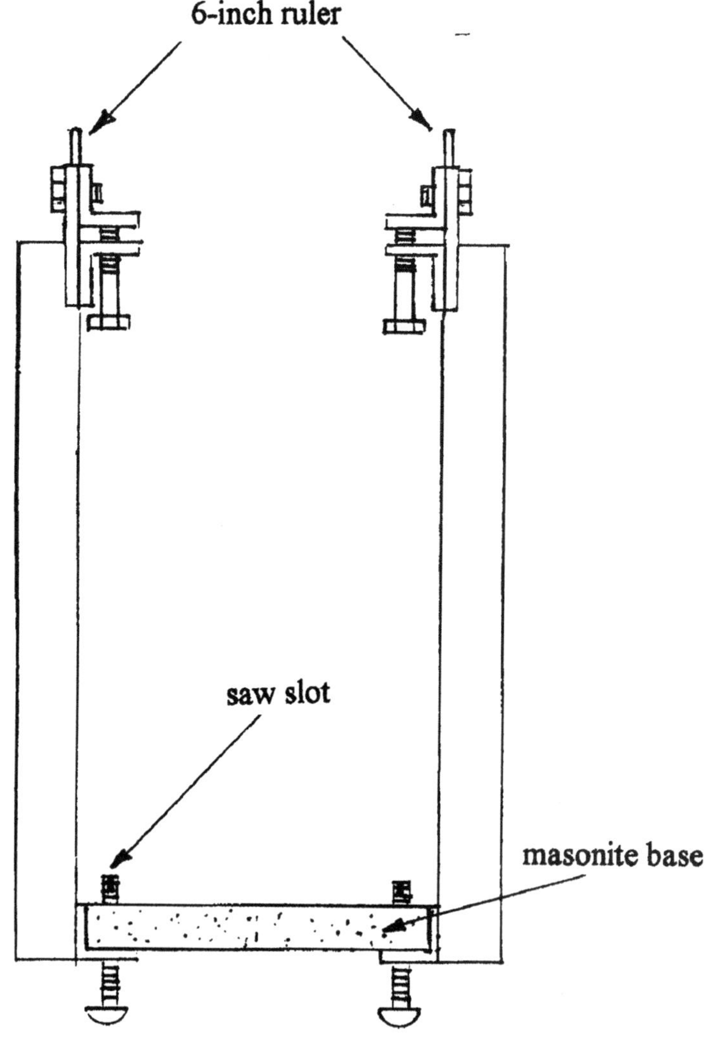

Balancing Ways End View

Shop made Balancing Ways

SCROLL HOUSING

Layout of the fan scroll is done using the offset and three radii calculated in the fan designing section. For a 10-inch wheel the offset is 1-inch, the three radii are 5-inches, 7-inches, and 9-inches, the cutoff point is 1/2-inch. When doing the layout you must add 3/8 of an inch to each radii to account for the flange that will extend around the edge for joining the housing together. Welding is the preferred method of joining the scroll housing sides together; however, sheet-metal screws will work. To begin the layout, locate and center punch the center of the wheel on the scroll housing. Scribe the wheel around the center point. Scribe the horizontal and vertical center lines. Locate and center punch the offsets to the right and left of the vertical center line. The layout of the scroll housing is shown in the drawings below.

Lay out and drill or cut the holes for the shaft and the air inlet. Because the spinning fan would quickly remove a finger is it were stuck into the inlet, it would be a good idea to cut a section of hardware cloth and a mounting ring to cover the inlet. Hardware cloth is available at any hardware store.

After the scroll has been cut out, bending the 3/8-inch flange around the scroll and welded together, or it may be left straight and joined with screws. Welding is the preferred method. To form a flange for welding, bend the edge up approximately 1/3 of the way (30°) with a pair of pliers. Make a second pass, bending the flange up another 30°, make a final pass completing the bend. The flange will smooth quite well with a hammer and auto-body dolly. If no dollies are available, a short section of 3-inch round stock will work. After the scroll has been welded it is cleaned up with a grinder and given a coat of Bondo. The Bondo is used to fill any rough spots and dents so the cover of the fan will have a smooth surface on which to seat.

If you choose to use screws to join your housing, you should scribe a line 3/16-inch in from the scroll edge and locate your screw holes. The center strip that joins the two scrolls together is cut with ¾ -inch extra width. Two edges are bent up as was done on the scrolls. The strip is bent around the scroll and drilled for the screws. Using vise grip pliers to help hold the strip in place, drill a screw hole, insert a screw, move the pliers down and repeat the process until all the screws are in. When the assembly is completely finished, seal it with caulking or silicone sealant.

Generation of a scroll housing is show in the four steps illustrated below.

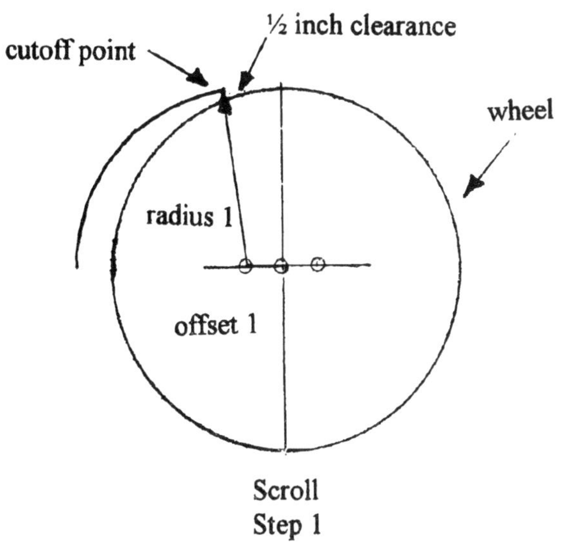

cutoff point
½ inch clearance
wheel
radius 1
offset 1
Scroll
Step 1

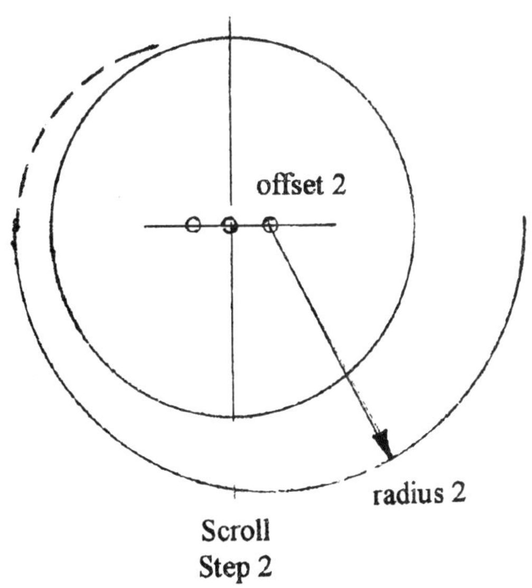

offset 2
radius 2
Scroll
Step 2

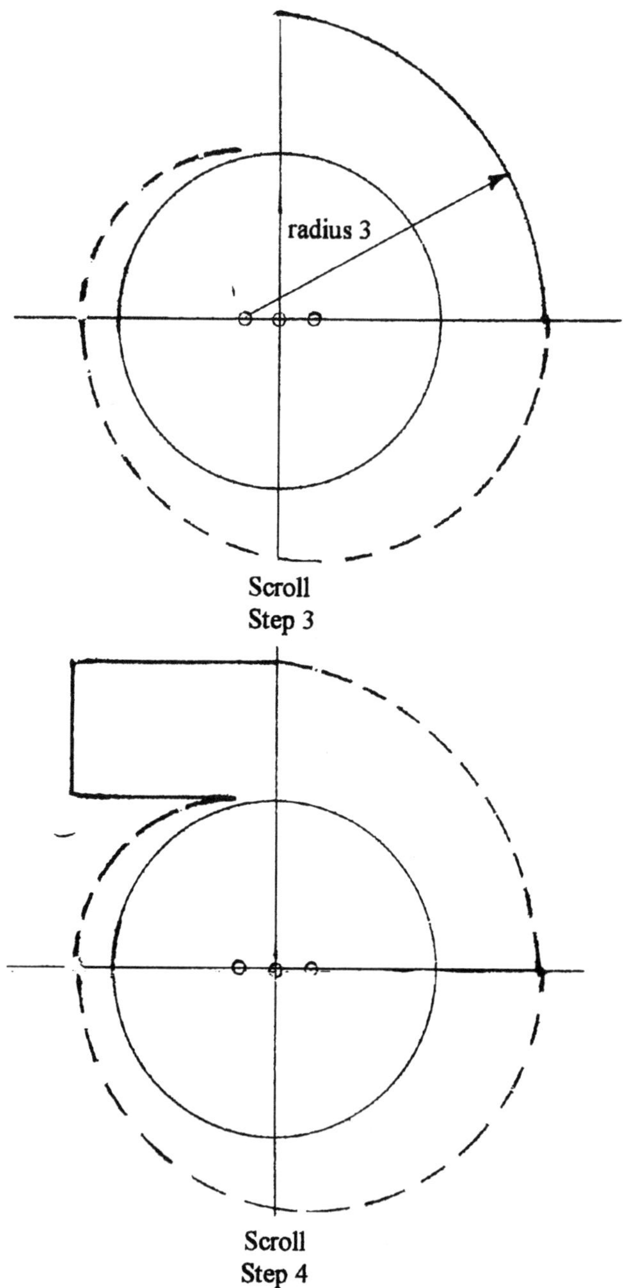

radius 3
Scroll
Step 3
Scroll
Step 4

VENTURI INLET:

The venturi inlet serves two purposes. First it increases the efficiency of the inlet from 15 to 28%. Second, it minimizes the recirculation of air within the fan. Tight clearance between the venturi and the inlet of the fan wheel will prevent the output air from being sucked back into the inlet.

Make a ring shaped pattern with the 2 ½ ° draft from a 2 x 4. Cut 6 sections with 60° sides, these will glue up to form a hexagon shaped ring. The rounded ring shape and the proper draft are cut in the lathe. The pattern is sanded smooth and painted.

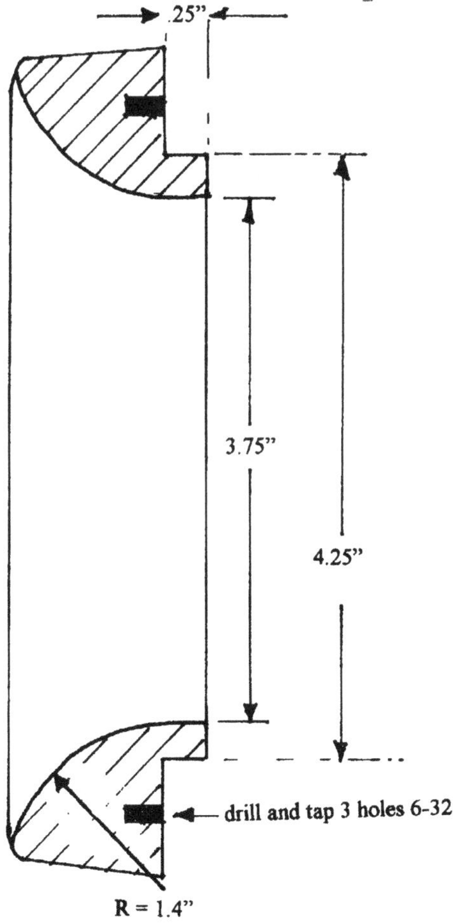

Pour the venturi inlet from aluminum. Finish the ring by turning in the lathe. The venturi is roughed out at a 45° angle. Make the final shape by forming with a wood turning chisel. Check the shape as you turn by using a cardboard template with the proper radius. The radius should be a minimum of 14% of the wheel diameter. This is 1.4-inches for the 10-inch fan wheel. Sand and polish the venturi after the radius is formed. This is a quick job in the lathe. Three holes are located 120^oapart on the backside of the venturi. The ring is drilled and tapped for 6-32 screws, which attach it to the blower housing.

TRANSITION PIECE:

The transition piece is also poured from aluminum. Make a split pattern and a core box. Both the pattern and the core box are sanded smooth and filled with Bondo autobody putty. The transitional radius on the inside of the core box is also formed with Bondo and sanded smooth. Cut the flange from ¼ inch thick plywood. The flange is given draft by building up Bondo and sanding the taper.

The core mix is made from sand, wheat paste (wallpaper paste) and molasses water. The water mixture is roughly 10% to 20% molasses and the balance is water. Mix a ratio of 90% sand to 10% wheat paste (more may be added if needed). Add the molasses water until the mass will hold a shape. Press it into the corebox and add two reinforcing wires. Flip the corebox over on a section of sheet steel for baking. Make two more cores (you will have an extra incase one breaks). Bake the cores at 350 for about a half an hour. The cores should be completely dry and brick hard. The cores are removed from the oven and glued together with a thick mixture of wheat paste and molasses water.

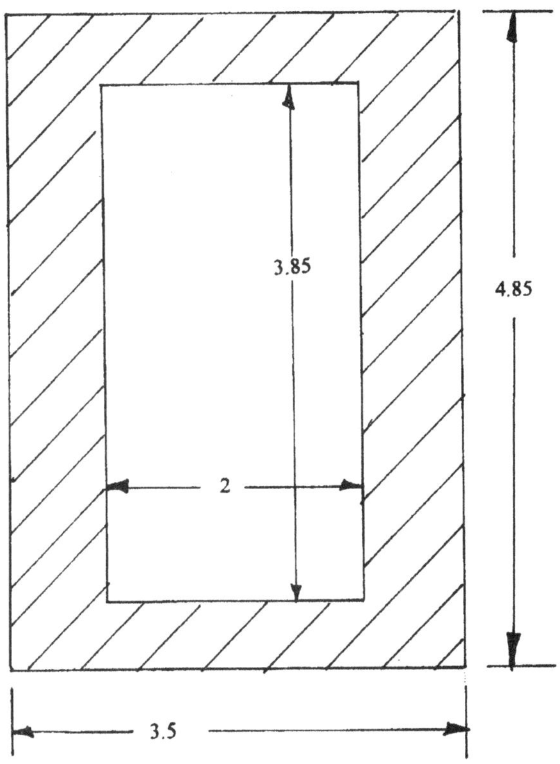

Fan End of Transition Piece

The core is placed in the mold and the mold is immediately poured. A core left in the mold will absorb water and make a gassy casting.

Clean up the casting and drill the flange with three equally spaced 9/32-inch holes. Later, the flange will be attached to the blast main with ¼ inch bolts. A 1/2-inch lip is bent out around the discharge opening of the fan scroll housing. Attach the casting to the fan with four 10/32-inch screws. Drill and tap the casting to accept the screws. Make a cork gasket to seal the area between the casting and the fan housing. Cork gasket material is readily available at auto part stores.

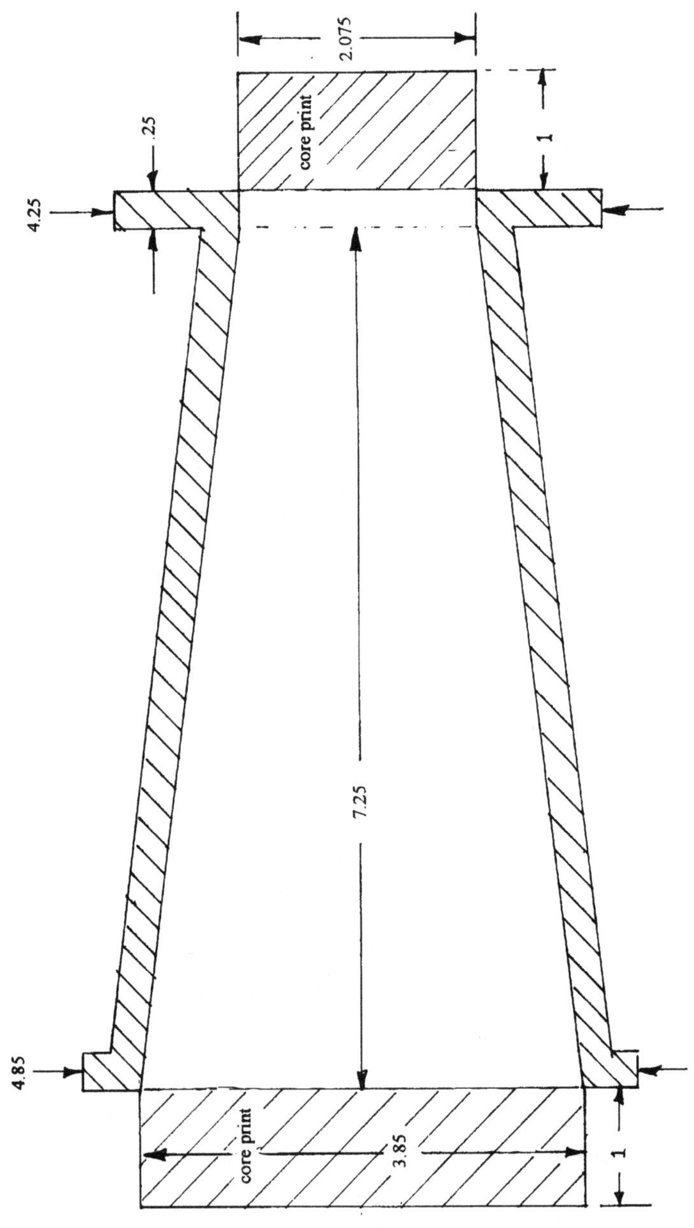

Layout of Flanged Transition Piece

Split Pattern and Core Box

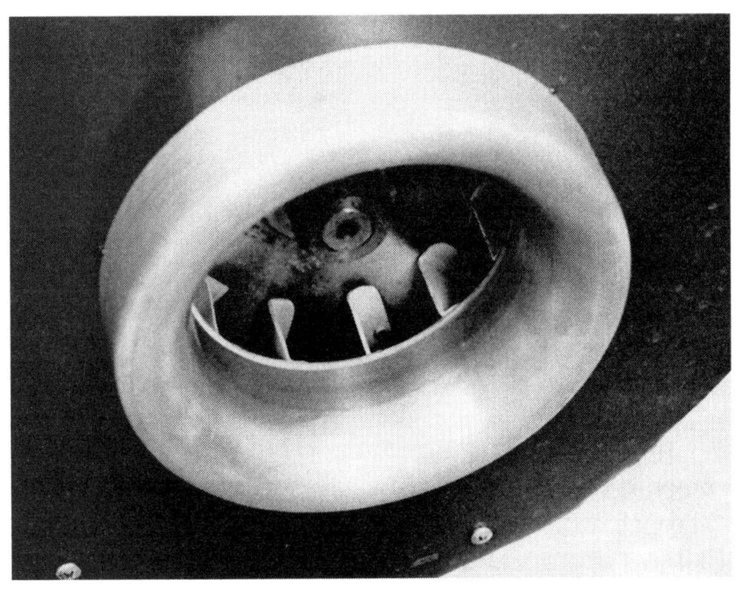

Tight Clearance of the Venturi Inlet

Completed Cupola Blower

If properly constructed, these blowers move a considerable amount of air. You will be pleasantly surprised by the strong blast they produce. I doubled the output of my cupola by building a proper blower.

VII. CONSTRUCTION OF PITOT TUBE AND MANOMETER.

Construction of a pitot tube is simple and should not take more than an afternoon. Start with a section of soft 3/8-inch diameter copper tubing. Carefully drill the eight holes as shown in the drawing. Remove any burrs from the cut ends and attach the Tee coupling using electronic solder and flux (available at radio shack). Be sure the end is bright and clean before you solder. A small propane torch makes fast work of this job. Make the nosepiece, the two brass plugs, and the spacer on the lathe. Cut a section of 1/8 inch tubing for the center tube. This tubing is used for automotive oil pressure lines, and can be found at auto parts stores. <u>Carefully measure the distance for the spacer</u> and solder into position on the inner tube. Solder the nosepiece to the inner tube and insert the assembly into the 3/8-inch tubing. Solder the nose to the 3/8-inch tubing. Do not solder the spacer to the 3/8 tubing. Fit and solder the remaining brass plugs in position. Polish the nose piece and tube brightly. If you are using a buffing wheel, it is much easier to polish the tube before bending. Using a tubing bender carefully bend the assembly. Apply a coat of clear lacquer. Although our tube does not meet laboratory specifications it will still work suprisingly well. I tested my tube on a still day by riding down the street and recording the inches pressure vs. speedometer. It was accurate to ½ mph.

The *Air Movement and Control Association* (AMCA) states the pitot tube stem diameter should not exceed 1/30 the test duct diameter. While our tube may not meet this specification, it works well enough for our purposes.

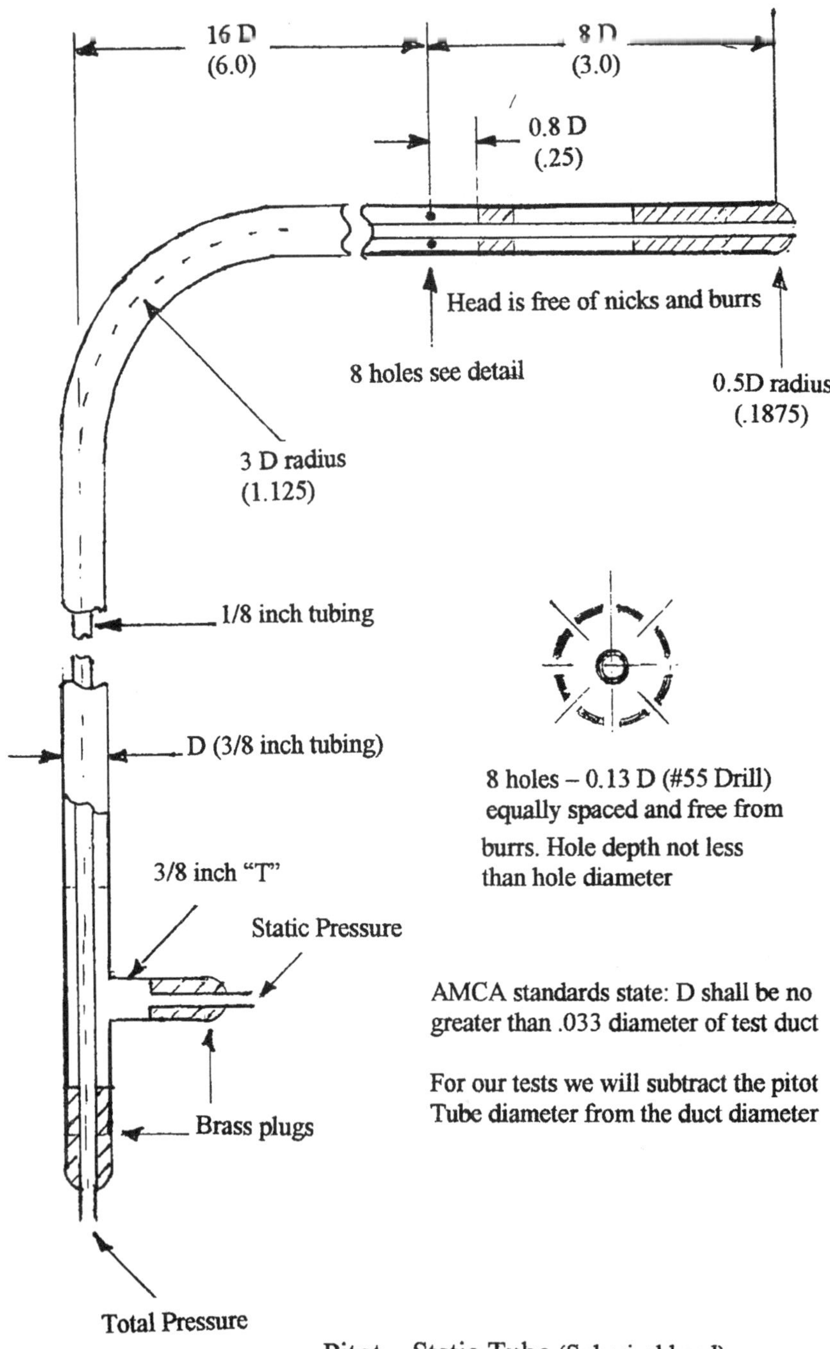

Pitot – Static Tube (Spherical head)

CONSTRUCTION OF THE MANOMETER.

Construction of the manometer should only take an afternoon. Use a flex rule for the manometer scale. Make the manometer frame from wood or aluminum. The manometer seen in the photo of the cupola is constructed of two sheets of ¼ inch aluminum plate covered with a sheet of Plexiglas. The section of Plexiglas is so small it can often be found as a scrap at the local hardware store.

Screw one aluminum plate to a piece of wood, and mill a groove for the ruler that is .010 deep and as wide as your ruler. Mount the form and wooden base to a rotary table and mill the "U"groove. Drill and tap the screw holes for the complete assembly. The center section of the U form is held in place from the rear by 6-32 screws. Assemble the frame (back section that has only screw holes and two pieces of U form) and Plexiglas cover. Trim the assembly to size on the mill. Smooth the edges with a sander. Open the form, press the tube into position, and replace the Plexiglas cover. Put a slight bend in the flex-rule and insert into its groove.

Alternately the frame could be cut from ¼ plywood. Carefully trim the semicircular section as accurately as possible. Rout a shallow groove for the ruler then form the inside section maintaining as smooth a curve as possible. Paint the form white for visibility. Assemble the form, press the tubing into position and attach the Plexiglas cover. The tension of the rule can be adjusted by slightly bending the flex-rule.

The manometer is half filled with a dilute solution of food coloring and water.   Dimensions are given for the manometer used on the 10-inch cupola. These are not critical, build it according to materials on hand. Just keep the tubes straight, and the curved section as smooth as possible. If you are going to use a higher pressure, you will need a longer manometer, or use a liquid that is heavier than water.

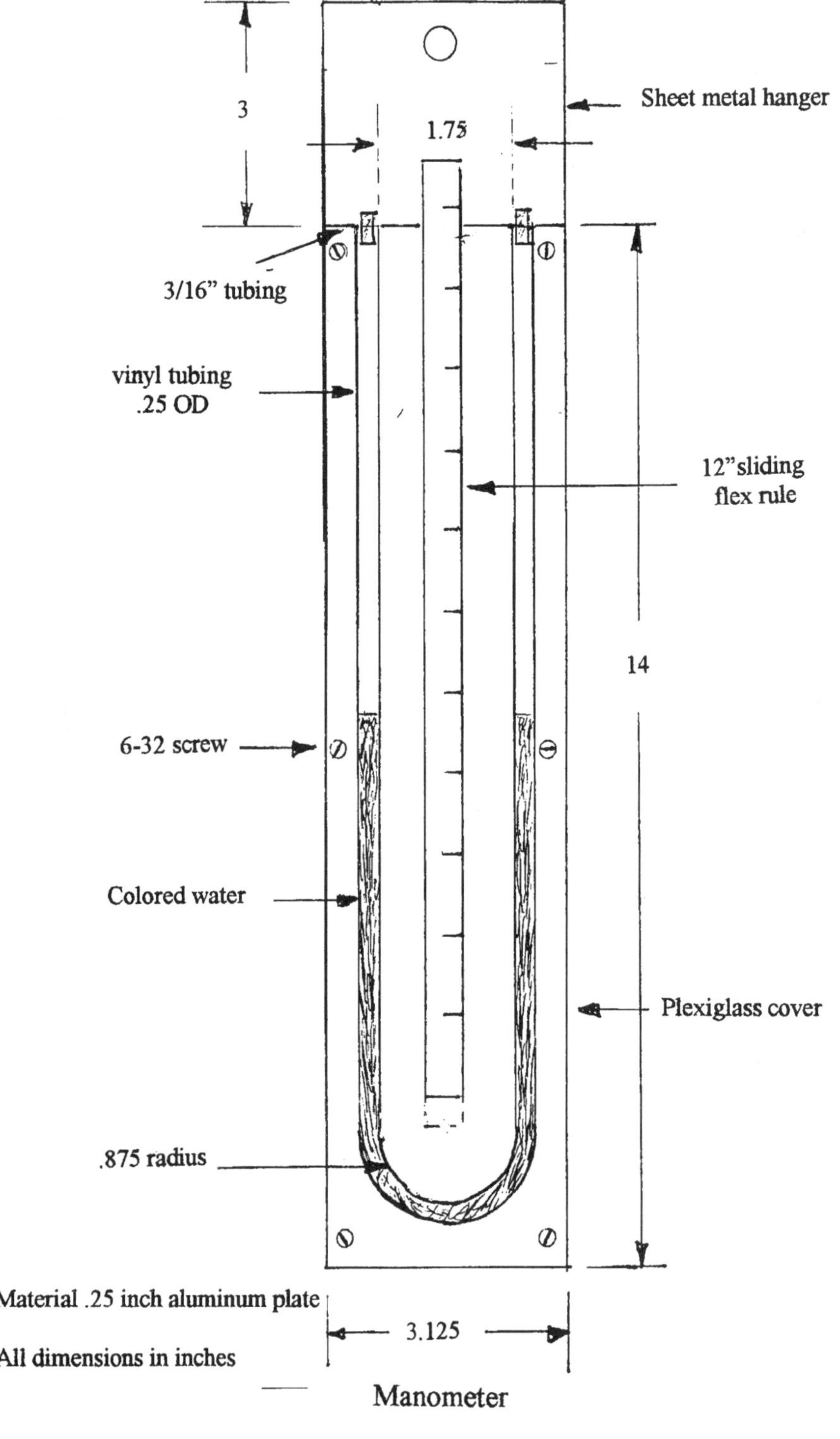

Sheet metal hanger
3
1.75
3/16" tubing
vinyl tubing
.25 OD
12"sliding
flex rule
14
6-32 screw
Colored water
Plexiglass cover
.875 radius
Material .25 inch aluminum plate
3.125
All dimensions in inches
Manometer

Calculate the new "inches-rise / fpm" of the new material according to the specific gravity (density of the new liquid relative to water).

Calculate the new scale for a manometer using mercury:

Specific Gravity of Mercury = 13.6 gm/cm^3

Specific Gravity of Water = 1 gm/cm^3

Inches rise = (1.73 inches) x (SG water / SG mercury)

Inches rise = 1.73 inches x (.0735) = .127 inches

Mercury might be used for a large cupola or a cupola with a hot blast system because a hot blast system often requires an additional 12 ounces of pressure above the cold blast pressure. Glycerin might be used for smaller cupolas. Furthermore, using a liquid that is lighter than water (lower specific gravity) will make a more sensitive scale. Mineral spirits combined with an inclined scale will make a very sensitive scale. Flammable liquids should only be used when testing a fan and are not to be used around an operating cupola.

SG of Glycerin   = 1.26 gm/cm^3
SG of Ethyl Alcohol =   .81gm/cm^3

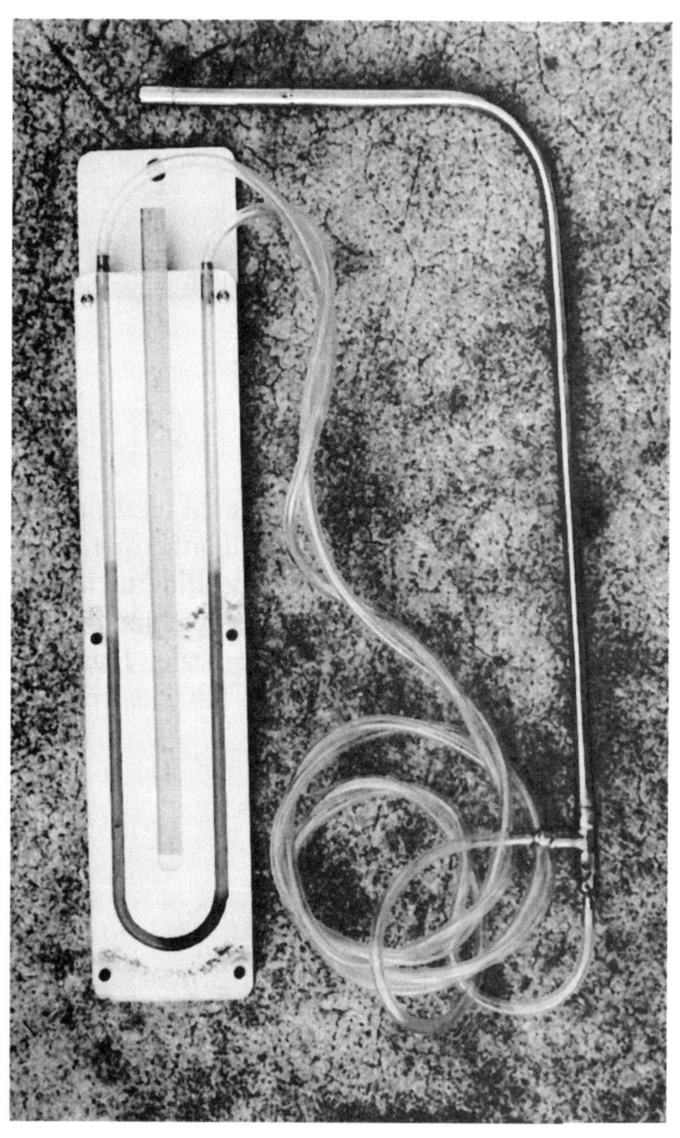

Shop made Pitot tube and manometer

VIII. CALCULATION OF AIRFLOW.

The optimal operation of the cupola requires the accurate measurement of the blast. A manometer attached to the windbelt will provide a good measure of cupola's operational condition. Such measurements, when recorded in a notebook, will provide a basis for future operation and help one achieve repeatable results. However, the actual volume of airflow must be measured with a Pitot tube.

Accurate measurement of airflow requires several readings across the diameter of the duct. Since our duct is so small in diameter, we will use one reading in the center of the duct (maximum velocity). Calculate the velocity at the center of the duct, and assume the average duct velocity to be 90 percent of this calculated maximum value. The actual velocity is lower than the maximum or center velocity because the air travels slower at the walls of the duct than in the center.

Turbulence in the airflow will not give accurate readings. Therefore the duct being tested should be smooth and free of burrs. Likewise the entrance and exits of the test duct should be smooth. Transition pieces or adapters that attach the fan to the test duct have no more than a 7 degree taper. By using a flow straightener and a duct that is ten times longer than it is wide you can attain minimal turbulence. L = 10D. A flow straightener, can be made from thin sheet metal, while it will not meet ACMA standards, it might be helpful.

The pitot tube is inserted through a hole in the side of the duct and pointed squarely into the airflow. Be sure to locate the hole in the duct so that the tip is 10 diameters from the duct input. The velocity of airflow will be calculated from the manometer reading.

Velocity of airflow is calculated by:

$$V = 4005\sqrt{h}$$

V = velocity in feet / minute
h = height in inches water column

Calculate the velocity of airflow for a manometer reading of 1.25 inches.

$$V = 4005\sqrt{1.25} \text{ inches}$$

$$V = 4478 \text{ feet per minute}$$

Volume of airflow is calculated by:

$$CFM = A \times V$$

$$CFM = \text{Area duct} \times \text{Velocity of airflow}$$

Calculate the volume of airflow for 10-inch cupola with a 2-inch diameter blast main:

Actual inside diameter of blast main $D = 2.067$ inches

Area of blast main in square feet $= .0218(D / 2)^2$

$$A = .0218(2.067 / 2)^2 = .0233 \text{ foot}^2$$

$$CFM = A \times V$$

$$CFM = (4478 \text{ ft} / \text{min}) \times (.0233 \text{ ft}) = 104.3$$

Average airflow is 90% the maximum calculated rate.

$$\text{Average airflow} = (.9) \times (104.3) = 93.9 \text{ CFM}$$

A more accurate calculation would subtract the area of the pitot from the test duct.

Area of pitot $= .000767 \text{ foot}^2$

The same calculations are performed with the new duct area the final flow being: 90.9 CFM

The airflow is a little low for best output; 125 to 165 CFM would be closer to commercial practice.

The effect of temperature on weight of air calculations: As the temperature of the surrounding air increases, the density of the air decreases. Hot air is lighter then cold air. This is obvious as hot air balloons rise. Our calculations are based on 70 degrees ambient temperature. The weight of 1 cubic foot of air at 70 degrees (sea level) is .0749 pounds. Air at 0 degrees is 15% heavier; likewise air at 100 degrees is about 6% lighter. Air at 400 degrees is about 40% lighter. If you wish to correct for air temperature in your calculations the table of air densities is listed below.

Density of air at 14.7 psia

Temperature degrees F	Density
0	.0863
32	.0807
42	.0791
52	.0776
62	.0760
70	.0749
80	.0735
90	.0722
100	.0709
200	.0602
400	.0461
600	.0400
800	.0337
1000	.0290

The fan output graph shown below is calculated for a 12 blade, 10" wheel, 1.1" wide, running at 5800 rpm.

Fan Test Duct with Two Manometers

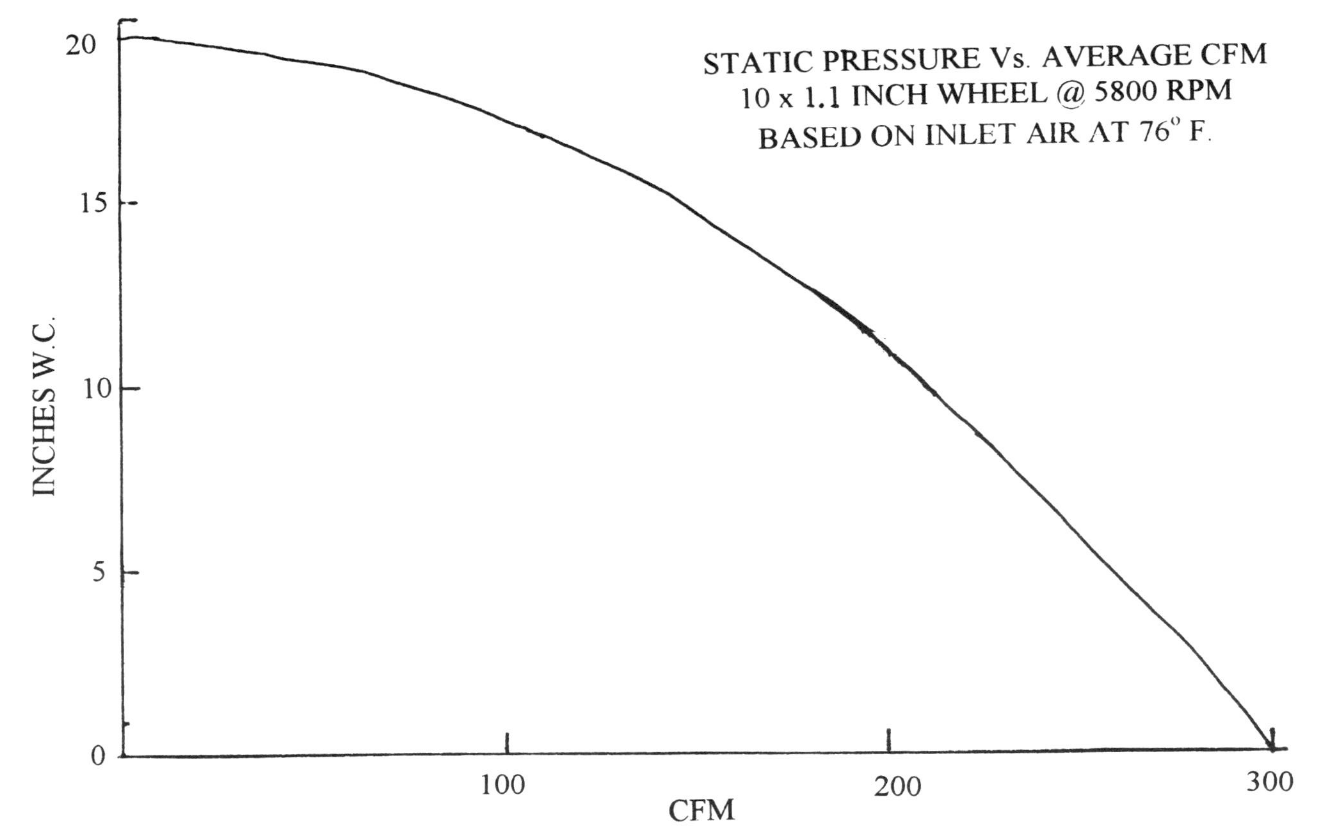

STATIC PRESSURE Vs. AVERAGE CFM
10 x 1.1 INCH WHEEL @ 5800 RPM
BASED ON INLET AIR AT 76° F.
INCHES W.C.
20
15
10
5
0
100
200
300
CFM

As you can see, the fan produces 300 cfm at 0 static pressure. It produces 19 inches water column at 0 cfm. At our target pressure of 13.84 inches water column it produces 165 cfm. Which is pretty much exactly what we are looking for. This is a powerful fan, be careful that the cupola is not overblown. Pay close attention to your manometers. Know your target values ahead of time. You will need to throttle this fan.

As stated early in the fan design section, actual output must be tested and the formula adjusted to reach desired output. The formulas predict the output will be 300 cfm, which is accurate at 0 static pressure. However fan output is a function of both static pressure and cfm. The adjusted formula that represents the output of 165 cfm at 13.84 inches water column is:

Width of wheel = (747 x cfm) / (D inches x tipspeed feet/min)

All other formulas remain the same.

You may use this new formula in your calculations, however it may be easier, now that the fan has been tested and graphed, to scale this fan up or down using the "fan laws" to meet the needs of any sized cupola.

IX. ADDITIONAL CUPOLAS BASED ON THE 15 INCH PROPANE CYLINDER SHELL:

Both larger and smaller cupolas may be built using the 15-inch diameter propane cylinder. The construction techniques are the same as described earlier. The main difference is the thickness of the refractory and the length of the wind belt. The refractory is thinned to 1.5 inches. Because of this, the inner cylinder will probably glow at red heat. Much heat will be lost through this wall, however the cooling effect of the blast should keep the cylinder from becoming too hot. One advantage of using thinner refractory will be considerable preheat of the blast. Actually most of the heat lost through the walls will be recycled back into the cupola. The thinner refractory will also lower the cost of building the unit.

Another difference is the bolt on well. Several wells, each of different capacity may be made for cupola. The 7-inch cupola has a 10" diameter sleeve of 16-gauge steel that holds the refractory. This may be difficult to roll; however the local technical collage may be able to do it for you. Thinner metal may be experimented with. Maybe several 5 gallon metal buckets could be welded together to form the inner sleeve. A disk cut from 16 gauge is welded to the top of the cylinder to secure an airtight connection for the wind belt. A shop vac should provide ample blast for the 7-inch unit.

The 12-inch unit is built of several cylinders bolted together. The bottom cylinder is lined with refractory 1.5-inches thick. The wind belt is constructed as used on the 10" cupola. The difference is the belt is extended the full length of the bottom cylinder. The lining of the upper cylinder tapers from 10 inches diameter at the top to 12 inches diameter at the bottom. This is to prevent the charge materials from hanging up in the shaft as they expand during the preheat. With both cupolas sand backing is specified to cut down on the cost of the units. A taphole opening is made by forming a strip of 16 gauge around the wooden form similar to the one used to form the taphole void in the refractory. The strip is welded into place between the inner and outer shells to keep the sand from falling through the shell. An alternative would be to use castable refractory to fill the empty space between the inner and outer shells of the well. With the proper blast, a tap of 100 pounds or more is possible with this cupola.

Tapping the Cupola
Note protective gear includes welder's hat and mask, goggles, gloves, long sleeves, long pants, and safety shoes

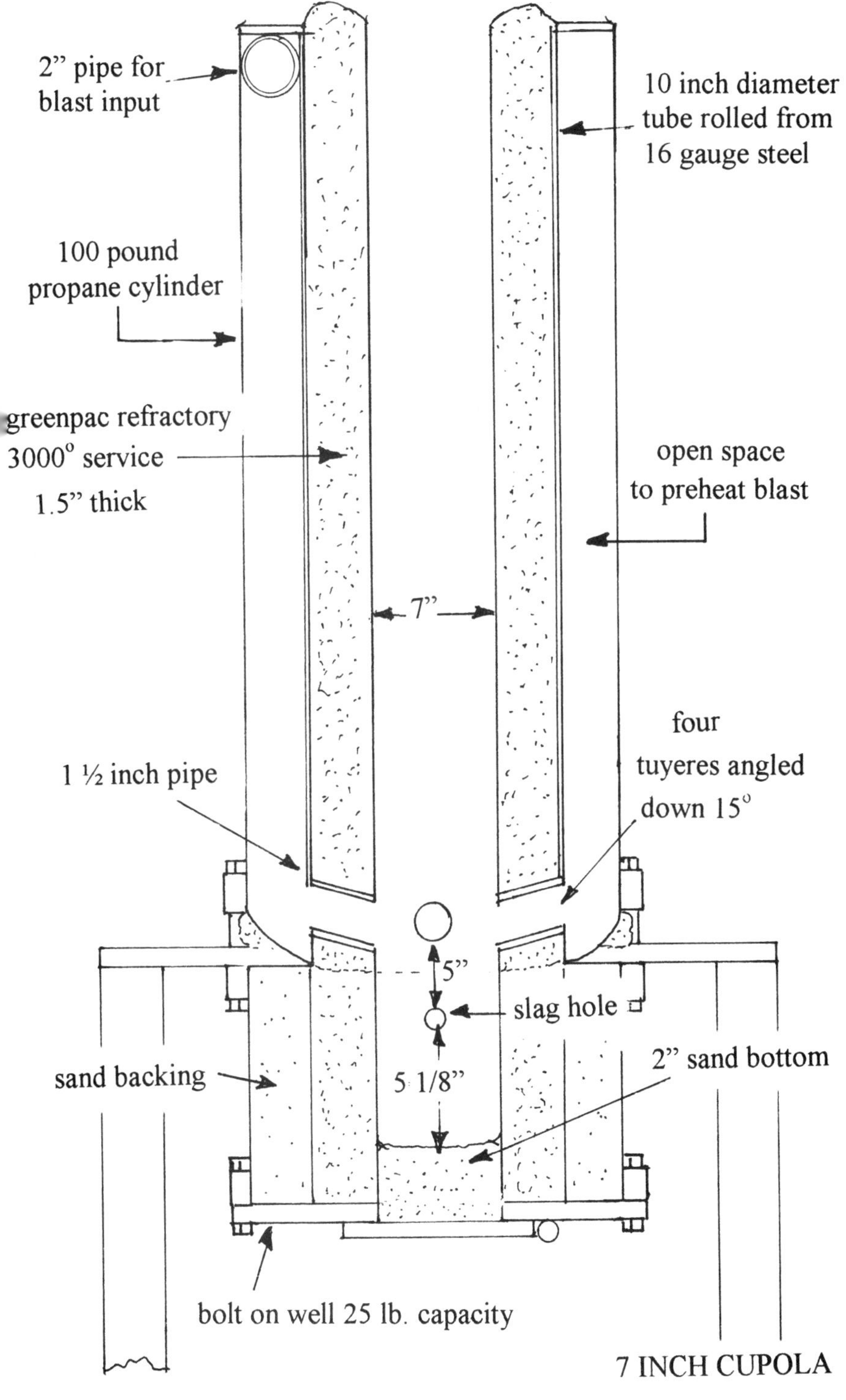

2" pipe for blast input
10 inch diameter tube rolled from 16 gauge steel
100 pound propane cylinder
greenpac refractory 3000° service 1.5" thick
open space to preheat blast
7"
1 ½ inch pipe
four tuyeres angled down 15°
5"
slag hole
5 1/8"
sand backing
2" sand bottom
bolt on well 25 lb. capacity
7 INCH CUPOLA

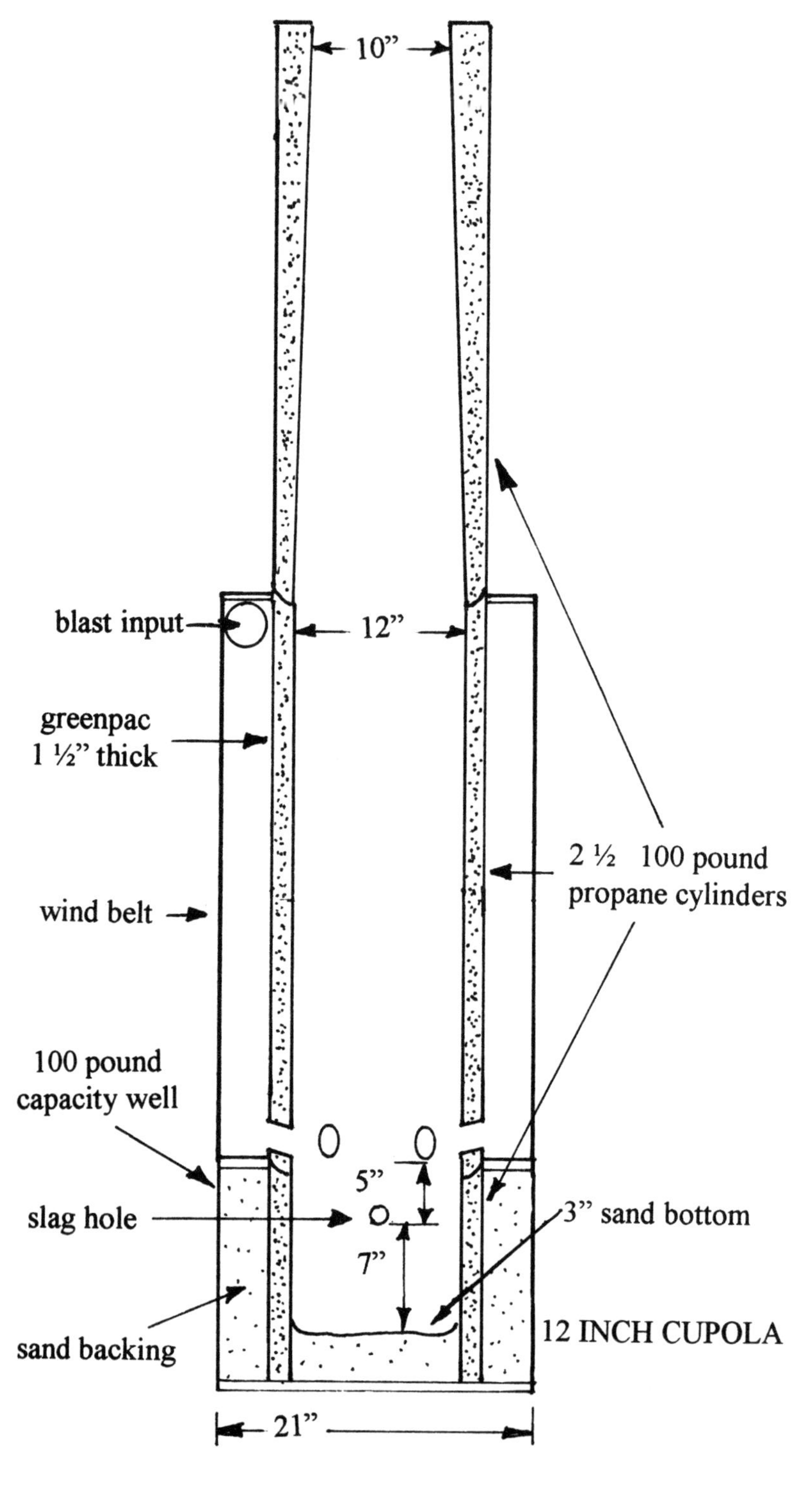

10"
blast input
12"
greenpac
1 ½" thick
2 ½ 100 pound
propane cylinders
wind belt
100 pound
capacity well
5"
slag hole
7"
3" sand bottom
sand backing
12 INCH CUPOLA
21"

X. OXYGEN ENRICHMENT:

Oxygen enrichment of the blast air is used for two main reasons, to increase the melt rate, or to increase the temperature of the tap. Other reasons for oxygen enrichment would include reduction of coke and the reduction of sulfur in the tap.

If a cupola is producing its maximum output, the output may be further increased by as much as 25% with the addition of oxygen to the blast air. Production increases relative to the percent higher oxygen levels are summarized below.

Oxygen	Increase in melt rate
1%	8%
2%	15%
3%	21%
4%	25%

TAP TEMPERATURE INCREASES:

If the blast is reduced and oxygen is introduced to the new lower blast volume, the temperature of the tap will increase. The increase is largely related to the increase in radiated heat of combustion. Radiated heat <u>increases</u> <u>rapidly</u> with an increase in temperature. The increase is on the order of the fourth power as illustrated below.

$$R = K(T_1^4 - T_2^4)$$

R = radiated heat K = a constant

T_1 = flame temperature T_2 = temperature of iron

Air is 21% oxygen and 79% nitrogen. The maximum temperature of combustion of coke in air is bout 3500° ; however if the percentage of oxygen is increased to 25% (4%

onrichmont) tho tomporaturo of oombustion oxooode 3800°. At 30% oxygen the temperature of combustion approaches 4200°. At such temperature the refractory disintegrates and the whole cupola comes apart. Careful calculation and monitoring of the airflow are critical to prevent a catastrophic failure of the cupola. A 4% increase in oxygen should increase the tap temperature about 175°. The reduction of the blast is dependent upon the percent of oxygen introduced and is summarized in the table below.

Oxygen	Blast Reduction
1%	6%
2%	11%
3%	16%
4%	20%

Calculate the cfm of oxygen for a 165-cfm blast:

Cfm oxygen =   cfm x (% oxygen / 100)

Cfm oxygen = 165 x (4 / 100) = 6.6 cfm

Calculate the new reduced blast volume for a 4% increase in oxygen if the current blast is 165 cfm:

New cfm = cfm – (cfm x (% reduction / 100))

New cfm = 165 - (165 x (20 / 100)) = 132 cfm

The tuyere diameter may have to be reduced at the lower volumes in order to increase the blast velocity to provide sufficient penetration of the bed coke.

Oxygen enrichment may be used on a continuous or an intermittent basis. For example if you wanted to reduce the amount of cold iron at startup you could use oxygen enrichment. Likewise, if you needed a particularly hot tap you could use oxygen enrichment.

COKE AND SULFUR REDUCTIONS:

Production of ductile iron requires very low levels of sulfur. Sulfur pickup is reduced by raising the temperature and by reducing the iron's exposure to sulfur contained in the coke. Foundry coke contains about .6% sulfur. A 2% oxygen enrichment with reduced blast enable one to reduce the charge coke by 15%, likewise a 15% reduction of sulfur is found in the iron.

METHODS OF OXYGEN INTRODUCTION:

Oxygen is introduced into the cupola by two methods, either by injection into the blast main or by injection into the tuyeres. Most enriched cupolas introduce the oxygen into the blast main through a diffuser. The use of preheated blast may require the use of tuyere injection.

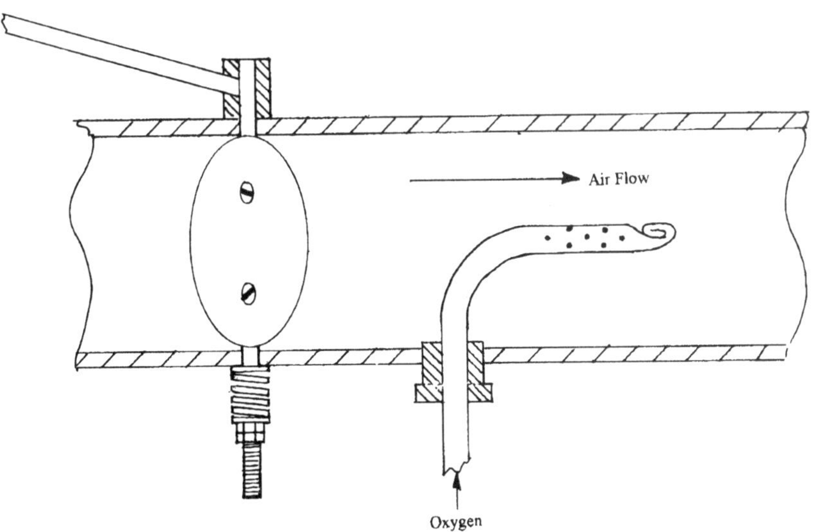

Oxygen Diffuser

Tuyere injection is accomplished by the use of an oxygen lance. The tip of the lance is kept about 3-inches from the end of the tuyere to keep it from burning up. Stainless steel is the preferred material for the lance used in tuyere injection.

The combination of hot blast and oxygen injection may raise the temperature of the tuyeres to the melting point. This is controlled by the use of water cooled tuyeres. Water-cooled tuyeres are made by tightly coiling copper tubing around the tuyere. Water is circulated through the assembly to keep it cool.

REGULATION OF OXYGEN ENRICHMENT:

Commercial operations use liquid oxygen in thermos bottles to supply their furnaces. We will use a welding cylinder, regulator, and a panel mount flow gauge. A suitable gauge is made by "King" and is available through MSC industrial supply. The current price for a 0-20-cfm gauge is $72; a smaller 0-4 cfm gauge is $42.

Because cupola operation is fast paced one would be well advised to calculate the desired inches pressure to be displayed on the Pitot tube and manometer. Likewise the oxygen flow rate should be calculated beforehand. A chart is made up and displayed beside the meters. Your manometer could be calibrated in cfm instead of inches to eliminate the volume of airflow calculations during melting.

XI. PURCHASE OF COKE:

Coke is sold in several sizes. Most large consumers purchase coke that is 3 inches in diameter and greater. Some suppliers sell smaller coke, often called "nut coke" that is 1½ inches and smaller in diameter. This size is easy to use in the small cupola. If you purchase the larger coke you will have to crush it to the correct size.

Coke is usually sold in train car quantities and larger. Remember this when dealing with your coke supplier, because he is often busy with several large orders. Most Suppliers will work you into their schedule, but be patient.

A ton of coke will fit nicely into a full sized pickup bed. I use a Chevy half-ton pickup and have no problems. It usually takes about two scoops from a front-end loader to fill the truck and the whole process takes about five minutes. I am currently paying 5 cents a pound, making the ton of coke $100. Some suppliers sell a ton bag. They dump the coke on a tarp, then tie it up at the top. It is put on a pallet and shipped in this fashion. Often the freight cost more than the coke. A ton of coke will give about 10 heats in the 10- inch cupola.

Coke may be found locally through the phone book, but I found my coke through the Internet. Another source of coke might be a local foundry.

I drive to Alabama to purchase my coke, as there are several coke companies located there. I found the people at Empire Coke the most accommodating; they even took me on a tour of the plant. Another popular company is ABC coke. There are several coke suppliers throughout the country and you may be fortunate enough to find one close to your location.

XII. CONCLUSION:

Cupola melting is not difficult. You can successfully melt without ever performing a single calculation. Do not be intimidated by the mathematical descriptions provided in the text. Chapters II, III, and IV certainly contain enough information to successfully melt iron; however if you want to design and operate your own cupola, the math is a very valuable tool that should provide you with good results. Remember every cupola will operate over a wide range of conditions. The calculations will put you in the ballpark but you will have to fine-tune your operating procedure by trial and error.

Your first cupola should be a small one, 10 to 12 inches or less. The temptation to build a large cupola is great, however you should resist until you have gained some experience on a small one. Because cupolas are such fast melters, you will have to have all your molds ready ahead of time. Preparing such a large quantity of sand is more than one person can easily handle. The 10-inch cupola, when operated with proper blast will melt 660 pounds of iron per hour. A 1 hour heat will require about 2 tons of sand. The amount of sand used for a large cupola can be overwhelming. My advice is to start small. You'll soon know what you can handle.

For a small operation that occasionally needs a large tap, a small cupola with a deep well and oxygen injection might provide you with the extra capacity required. Oxygen injection is for advanced and experienced cupola work and should not be attempted by the beginner.

Iron exists in many forms, gray, ductile, white, mottled, malleable, etc. You should read a book on basic metallurgy. Initially you might not understand it, but after inspection of the castings from your first few melts you will. Try putting a few charges of steel in the cupola and pour some test bars. Fracture the casting and you will see white iron. It is extremely hard and brittle. Your first few heats will be experimental so you can learn the process.

Operating a cupola is hard work, but there is nothing like the excitement of tapping white-hot iron and filling a mold. I certainly enjoy it and you will too. Now that you can produce iron castings you are only limited by your imagination. Good luck.

Appendix

SUPPLIERS:

American Foundrymen's Society
505 State Street
Des Plaines, IL 60016-8399

Foundry books, from basic to advanced
800 537 4237 www.afsinc.com
moderncastaing.com

COKE:

ABC Coke
800 321-4015
205 849 1300

Hickman Williams
US 800 862-1890
Canada 800 265 6415
Mexico   011 52 8 363-4041

Empire Coke
Holt, Alabama
205 323-2400
800 544 8702

ASi International
Cleveland , OH
800 860-4766
216 485-1800

Asbury Cast Metals Group
Sunbury, PA
717 286-2176

Miller & CO
Rosemont, IL
800 727-9847
847 696-2400

St. Louis Coke & Foundry Supply
314 727-7500

Joy-Mark Inc.
2121 E Norse Ave.
Cudahy, Wis. 53110
414 769-8155
Ceramic reinforced ladles.
Part# 407   10 x 8.5 x 11.375

Lindsay Publications   815 935 - 5353
P.O. Box 538   lindsaybks.com
Bradley, IL 60915   (Foundry books , Technical books)

MSC Industrial Supply Co.
75 Maxess Road
Melville, NY 11747-9415
(800) 645-7270

Request the "Big Book" catalog

Budget Casting Supply
San Mateo CA
650 345-3891   budgetcastingsupply.com

Piedmont Foundry Supplies
P.O. Box 87
3191 Rogers Lane   Columbus, GA Phone 324-5938
Cloverdale, VA 24077 Phone 992-3911,   Greensboro, NC, Phone 294-2661

Porter Warner Foundry Supplies:

Alabama   205 251-8223   Tennessee 423 266-4735
Arizona   602 244-9166
California 213 722-1335   TX   817 633-1767
           501 633 0285   713 222 6210
           909 822 5591

South Carolina 803 789-3444

Uni. West Foundry Supplies:

Denver ,CO   303 388-0922   Washington   206 767-9880

Oregon   503 226-4836

Rice Industries
Minnesota 651 784-1881

Stewart Marshall
P.O. Box 279
Lopez Island, WA 98261

"Building Small Cupola Furnaces" Book, Foundry CD.

Australian Foundry Institute
1 St. Peters way
Glengowrie, SA 5044
61 8 8295 67 12

Canadian Foundry Association
Ste. 1500
Ottawa, Ontario KIN 7B7
Canada

Cast Metals Institute   www.castmetals.com
800 537-4237

Average Airflow In CFM Through Pipe

		Diameter of Pipe					
Inches wc	Velocity ft/min	2 inches	2 1/2 inches	3 inches	3 1/2 inches	4 inches	5 inches
0.50	2831.96	56	85	131	175	225	423
0.75	3468.43	68	104	160	214	276	518
1.00	4005.00	79	120	185	247	319	598
1.25	4477.73	88	134	207	276	356	669
1.50	4905.10	96	147	226	303	390	733
1.75	5298.12	104	158	245	327	422	792
2.00	5663.93	111	169	262	350	451	846
2.25	6007.50	118	180	277	371	478	898
2.50	6332.46	124	189	292	391	504	946
2.75	6641.54	130	198	307	410	528	992
3.00	6936.86	136	207	320	428	552	1036
3.25	7220.12	142	216	333	446	574	1079
3.50	7492.67	147	224	346	463	596	1119
3.75	7755.65	152	232	358	479	617	1159
4.00	8010.00	157	239	370	495	637	1197
4.25	8256.52	162	247	381	510	657	1234
4.50	8495.89	167	254	392	525	676	1269
4.75	8728.70	171	261	403	539	694	1304
5.00	8955.45	176	268	413	553	712	1338
5.25	9176.61	180	274	424	567	730	1371
5.50	9392.56	184	281	434	580	747	1403
5.75	9603.65	188	287	443	593	764	1435
6.00	9810.21	192	293	453	606	781	1466

6.25	10012.50	196	299	462	618	797	1496
6.50	10210.79	200	305	471	630	812	1525
6.75	10405.30	204	311	480	642	828	1555
7.00	10596.23	208	317	489	654	843	1583
7.25	10783.79	212	322	498	666	858	1611
7.50	10968.14	215	328	506	677	873	1639
7.75	11149.45	219	333	515	688	887	1666
8.00	11327.85	222	338	523	699	901	1692
8.25	11503.49	226	344	531	710	915	1719
8.50	11676.48	229	349	539	721	929	1744
8.75	11846.95	232	354	547	731	943	1770
9.00	12015.00	236	359	555	742	956	1795
9.25	12180.73	239	364	562	752	969	1820
9.50	12344.24	242	369	570	762	982	1844
9.75	12505.61	245	374	577	772	995	1868
10.00	12664.92	248	378	585	782	1008	1892
10.25	12822.26	252	383	592	792	1020	1916
10.50	12977.68	255	388	599	801	1033	1939
10.75	13131.27	258	392	606	811	1045	1962
11.00	13283.08	261	397	613	820	1057	1984
11.25	13433.18	264	401	620	829	1069	2007
11.50	13581.62	266	406	627	839	1081	2029
11.75	13728.45	269	410	634	848	1092	2051
12.00	13873.73	272	415	641	857	1104	2073
12.25	14017.50	275	419	647	865	1115	2094
12.50	14159.81	278	423	654	874	1127	2115

Capacity of Cupola Wells (approximate) per Inch of Depth:

Weight of Iron = $.0992\ D^2$ D = diameter of well

Well Diameter in inches	7	10	12	14	16	18	20	24
Pounds of iron per inch of well depth	4.86	9.92	14.28	19.44	25.40	32.14	39.68	57.14

Less than 10-inches diameter: max well capacity is 60% theoretical tap (see pages 16 and 21). 10-inches and greater, max well capacity = theoretical tap 18-inches and greater max well capacity = 1.6 x theoretical tap.

Weight of Coke Charges: = $.0618\ D^2$ D = diameter at tuyeres

Cupola Dimensions in inches:

Cupola Diameter	7	10	12	14	16	18	20	24
Well Depth*	.5 to .6	.5 to 1	.5 to 1	.5 to 1	.6 to 1.2	.6 to 1.6	1 to 1.6	1 to 1.6
Area sq. in	38.5	78.5	113	154	200	254	314	452
Constant	5	5	5 to 6	5 to 6	5 to 7	5 to 7.5	7.5	10
Tuyere Diameter	1.5	1.5 to 2	2	2	2	2.5	2.5	2.5
Number of Tuyeres	4	4	6	8	10	9 to 10	10	12
Stack Height min.	31.5	45	54	63	72	81	90	108
Optimal Stack	44.45	63.5	76.2	90	102	114	127	152
Coke Weight lbs	3	6.18	9	12	16	20	25	35
Iron ratio	4 to 5	4 to 6	5 to 6	6	6 to 8	6 to 8	6 to 8	6 to 10

*Well capacity as a ratio of 1 lb. of iron per square inch of area. Example: Find the well for a 10-inch cupola using the .5 ratio. .5 x 78.5 = 39.25 lbs. of iron. Well depth = 39.25lbs / 9.92 lbs. per inch of depth.   *Well depth = 3.96 inches.*

Fan Capacity CFM (approximate) **for Cupola Furnaces:**

Fan Capacity = **1.96 D^2**   D= Diameter between tuyeres

Diameter in inches between Tuyeres	7	10	12	14	16	18	20	24
Approximate Fan Capacity, CFM	96	196	282	384	501	635	784	1129

See pages 61 – 63 for fan capacity

Fan Dimensions* inches:

Cupola Diameter	7	10	12	14	16	18	20	24
SP oz.	6	8	10.5	12	14	16	17.5	20.5
rpm	5700	5700	5700	5700	5700	5700	5700	5700
Wheel Diameter	8.75	10	11.5	12.375	13.375	14.3	15	16.125
Inlet diameter	2.82	3.58	4	4.475	4.89	5.3	5.68	6.41
Wheel Width	0.625	0.98	1	1.1	1.17	1.25	1.3	1.5
R1	4.375	5	5.75	6.1875	6.6875	7.15	7.5	8.0625
R2	6.125	7	8.05	8.66	9.3625	10	10.5	11.28
R3	7.875	9	10.35	11.138	12.038	12.87	13.5	14.513
Offset	0.875	1	1.15	1.24	1.34	1.43	1.5	1.61
Cut off	0.4375	0.5	0.575	0.6188	0.6688	0.715	0.75	0.8063
Discharge area	8.2	14.7	17.25	20.42	23.47	26.81	29.25	36.28
hp	0.375	0.853	1.61	2.5	3.82	5.5	7.5	12.6

*Suggested dimensions. Many variations are possible. See pages 67 –76 for complete design data.

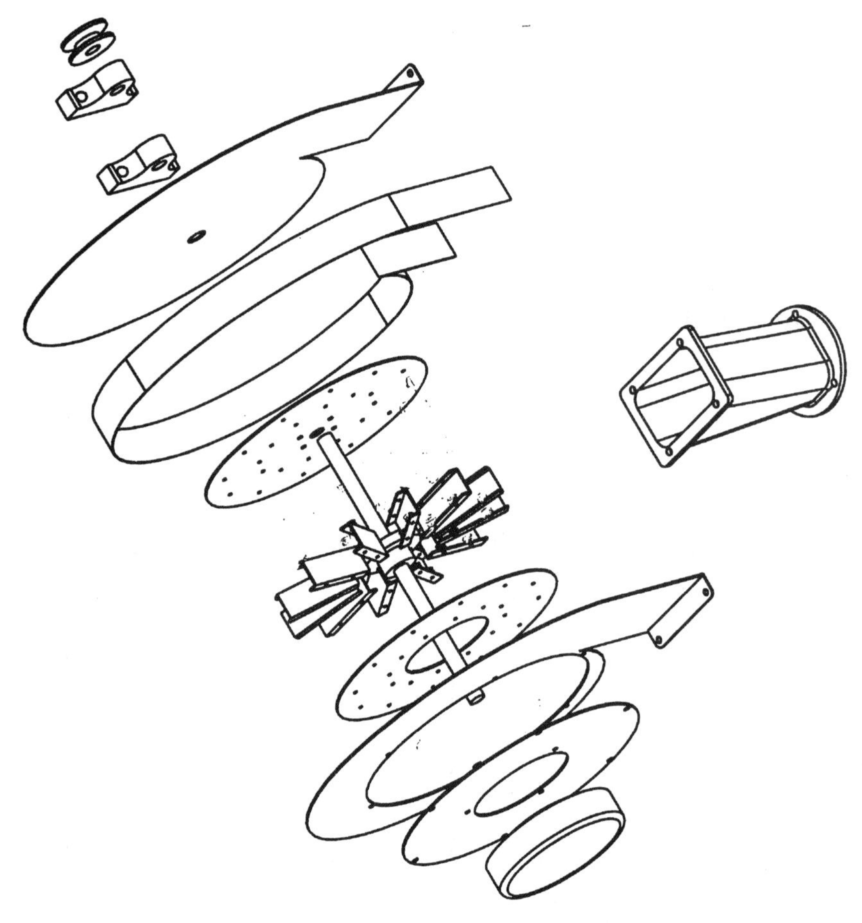

Fan Assembly: Exploded View

<u>**The Small Foundry Series :**</u>

Metal Casting: A Sand Casting Manual Vol. 1 & 2

Learn how to cast metal in sand molds using simple techniques and readily available materials. Steve Chastain, a Mechanical and Materials Engineer shows the beginner how to make a sand mold and then how to hone your skills to produce high quality castings. Written in non-technical terms, the sand casting manuals begin by melting aluminum cans over a charcoal fire and end by casting a cylinder head. All for little cost in your own back yard! Includes the basic metallurgy of aluminum, iron and copper-based alloys. Sold in over 30 countries, Chastain's popular "Small Foundry Series" is good for both the beginner and experienced metal caster.

Volume I   ISBN 0-9702203-2-4   208 Pages $19.95

Volume II   ISBN 0-9702203-3-4   192 Pages $19.95

Build a Tilting Furnace

Complete plans and operating instructions for a tilting furnace that easily melts 100 pounds of aluminum per hour. Melt with propane, diesel, or used motor oil! Using the tilting mechanism, you will never have to handle a hot crucible again. Start, stop, and hold the furnace at any angle for precise pours. Furnace may be modified to melt other metals.
ISBN 09702203-1-6   192 Pages $19.95

Making Pistons for Restoration Engines

Bring even the most "impossible" old engines back to life for little cost! You are no longer limited by the price and availability of replacement pistons and rings when you can make your own. Use inexpensive modern piston rings on your antique equipment! Design and make pistons for new or old engines. Learn to make all the tools and jigs needed to quickly produce top quality replacements in your own back yard. Heavily illustrated. A "must have" for antique equipment restorers!
ISBN 09702203-4-0   64 Pages $11.50

Iron Melting Cupola Furnaces

Complete plans and operating instructions for a 10" diameter cupola that will melt 330 pounds of iron per hour when powered by a shop vac! Also included are plans for a high-pressure blower that will increase the output of the small furnace to 660 pounds per hour. All can be built for little cost, mostly from scrap.
ISBN 0-9702203-0-8   128 pages $19.95

Quick Order Form

Email: stevechastain@hotmail.com

Credit card orders accepted through "paypal.com"

Postal Orders: Steve Chastain
2925 Mandarin Meadows Dr.
Jacksonville, FL 32223

Please send me the following books. I understand I may return them for a full refund:

NAME: ____

ADDRESS: ______________________________________

CITY: _______________________ STATE: __________

ZIP: __________

Number	Title	
______	IRON MELTING CUPOLA FURNACES FOR THE SMALL FOUNDRY.	$19.95
______	BUILD A TILTING FURNACE.	$19.95
______	A SAND CASTING MANUAL FOR THE SMALL FOUNDRY. **VOLUME 1**	$19.95
______	A SAND CASTING MANUAL FOR THE SMALL FOUNDRY. **VOLUME 2**	$19.95
______	MAKING PISTONS	$11.50

sub total: __________

Shipping: $1.75 for the first book, .75
for each additional book. __________

Total Enclosed: __________